Architect's FRAME 02

House, Searching for Life-Forms
Kim Seunghoy

주택, 삶의 형식을 찾아서
김승회

House, Searching for Life-Forms

Kim Seunghoy

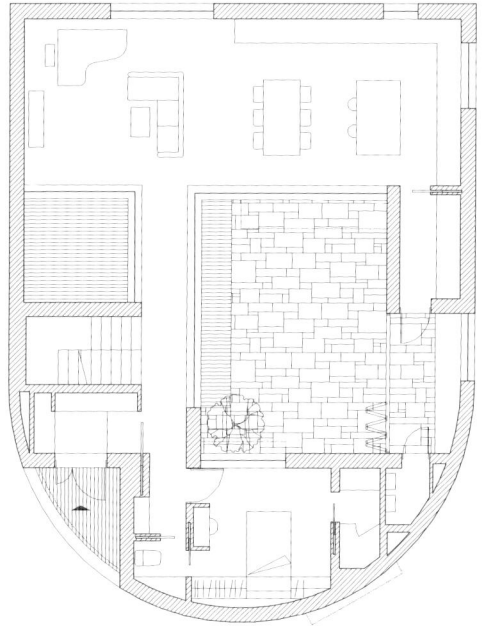

공간
서가

006 프롤로그

유형의 탐구

012 채와 간, 그리고 단일매스-박스
015 채와 간의 유형
028 단일매스-박스 유형

유형을 넘어서

056 탐구의 다양한 주제들
057 신도시의 주택들: 공공성과 개인의 취향
070 전원주택의 실험: 마당과 방, 그 소우주의 풍경
084 타운하우스와 마을-주거: 집합의 문법 찾기

주택의 변화를 부르는 것들

106 새로운 실험과 진화하는 주택

130 에필로그

006   Prologue

The Exploration of Types

012   *Chae* and *Kan* and the Single Unit Mass – Box
015   The Type of *Chae* and *Kan*
028   Single Mass – Box

Beyond Types

056   Various Topics of Exploration
057   Housing in New-Town Areas: Communality and Individuals
070   The Country House Experiment: Courtyard, Garden and Room, A Microcosmic Landscape
084   Town Houses and Village Residences: Looking for the Grammer of Collectivity

That which Transforms Houses

106   New Experiment and Evolving Housing

130   Epilogue

Prologue

1994년 독립된 건축가로 활동을 시작한 이후, 다양한 종류의 건축을 실험할 수 있는 기회를 가졌다. 학교, 병원, 교회, 업무시설, 보건소 등 몇몇 프로그램 – 유형은 연작의 방식으로 작업이 이어졌다. 그 과정을 통해 각 프로그램이 요구하는 공간 조직을 보다 깊이 탐색할 수 있었다. 특별히 지난 23년간 45개의 프로젝트로 꾸준히 이어져온 주택 연작은 내 작업의 변화와 생각을 가장 잘 드러내주는 길쭉한 단면도가 되었다.

처음 건축을 배우기 시작했을 때부터 지금까지, 주택 프로젝트는 언제나 건축에 대한 입장과 실천의 바탕을 마련해주었다. 내 작업의 근원에 '주택'이라는 저수지가 있다면, 주택 외의 다른 작업들은 그곳으로부터 시작된 여러 개의 물길이라고 할 수 있다. 그 까닭은 주택이라는 형식이 우리 '삶'의 진실에 가장 가깝고, 그 속에 깃든 삶은 건축이 필요한 절실한 이유가 되기 때문이다. 주택에 대한 생각은 학교와 병원, 교회, 업무시설 설계에 큰 영향을 주었다.

주택을 설계한다는 것은 가장 근원적인 건축의 형식을 탐구하는 일이다. 주택이라는 공간을 매개로 한 개인이 속한 시대와 장소와 대면하면서 삶과 세계의 총체성을 성찰한다. 주택은 건축가에게 건축과 인간, 사회와 지리에 대한 질문을 던지게 한다. 그 많은 질문을 지나서 마침내 하나의 주택이 태어난다.

건축에 대한 질문만으로는 건축이 만들어지지 않는다. 질문을 넘어서야 건축이 된다. 건축은 건축만의 고유한 논리와 형식이 있기 때문이다. 그럼에도, 건축이 물리적 실체인 동시에 사유를 내포하는 존재라면 건축의 바탕에 있었던 사유와 질문에 대해 피력하는 것 역시 의미 있는 일이라 생각한다. 건축가가 자신이 작업한 건축에 대해 문장으로 길게 적는 일은 분명 부끄러운 일이다. 그 부끄러움을 이기게 하는 것은 설계하면서 대면했던 생각과 질문, 그것의 절박한 기억이다. 그 기억을 바탕으로, 그간 진화를 거듭해왔던 주택 연작에 대한 생각을 몇 갈래로 나누어 밝히고자 한다.

In 1994, I began my own practice, beginning a number of diverse experiments across architectural types—schools, hospitals, churches, offices, and public clinics—with each project leading to another in the manner of a series. Through this process I was able to more deeply examine the spatial requirements of each programme. In particular, by considering the 45 residential projects from the past 23 years as a series provides a cross-sectional view of my work, and best illustrates the evolution in my thinking and in my work.

When I first began studying architecture, residential projects formed my perspective and provided context for my practice, as it continues to do so. If, at the core of my work lies the idea of the 'house' as a reservoir, one might say that my other projects are like waterways leading off from that reservoir. The reason for this is because the idea of a house or of a home is closest to the truth of what our lives are, and because architecture is so essentially necessary to our lives. My thoughts about what constitutes a house is greatly influenced my construction work on schools, hospitals, churches and offices.

To construct a house is to explore the most archetypal form of architecture. With a residential space as mediator, one can face the time and place in which one exists, reflect on life and on the totality of the world. To the architect, the house asks questions of architecture and man, society and geography. It is only after exploring many such questions that a house is born.

Architecture is not built with questions alone. One must go beyond questions to form architecture. This is because architecture has its own logic and form. In spite of this, if architecture is a physical reality while also connoting an architect's vision, it seems meaningful to express my thinking process as it relates to architecture. In truth, it is unbecoming for an architect to explicate at length on his own works. It is the urgent memory of the thoughts and questions I had when faced with construction that motivates me to make the attempt. I hope to illuminate the evolution of my residential projects by separating them into different strands.

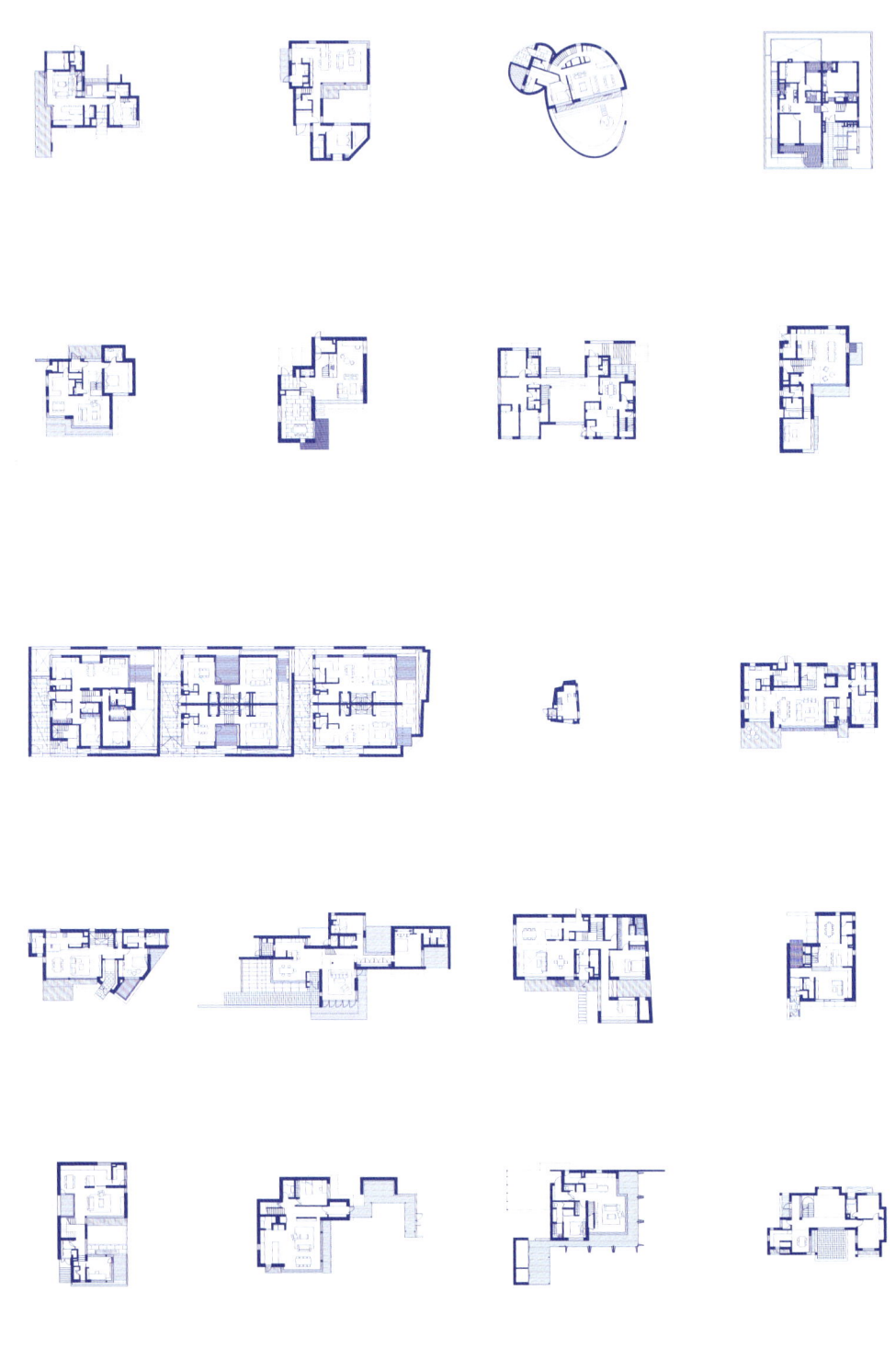

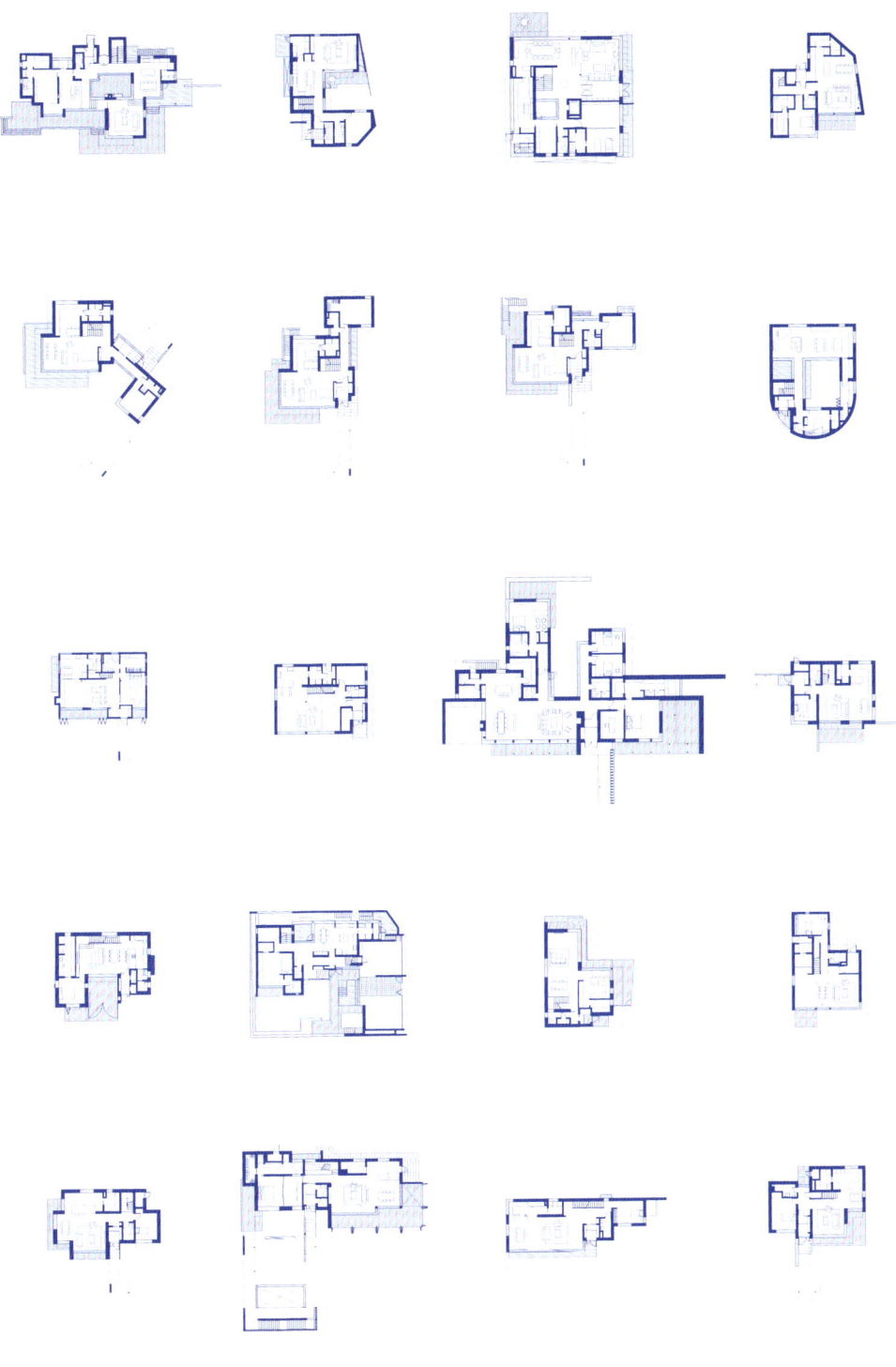

유형의 탐구

The Exploration of Types

채와 간, 그리고 단일매스-박스

건축을 배우던 수련기간 내내, 건축의 여러 관점과 이론들이 서로 다투었다. 모더니즘, 구조주의, 컨텍스추얼리즘, 타이플로지, 포스트모더니즘, 레이트모더니즘, 디컨스럭티비즘, 더치모던, 네오어버니즘 등, 흥미롭고 가치 있는 건축 이론이었지만, 나의 것으로 온전히 받아들여지지 않았다. 하나의 세계로 내게 다가오는 대신, 파편으로 쌓이거나 바람처럼 스쳐갔다. 건축가가 자기의 목소리로 설계를 시작할 때 반드시 필요한 것이 세계와 건축에 대한 어떤 입장이라면, 치열한 현실의 삶과 대면하는 시간을 견뎌야 비로소 그 입장이 현실의 기초 위에 세워지리라 믿었다.

Fig. 1 가회동 11번지 실측평면도 / 우리 시대 도시주택의 모델을 탐구하고 싶다는 바람은 대학원 시절 줄자를 갖고 가회동의 한옥들을 재고 다니던 때부터 키워온 것이다.
Gahoe-dong measured drawing / My desire to explore the models of modern urban residential housing is one I have nurtured since graduate school, when measuring the hanok buildings of Gahoe-dong in Byukchon.

    그 과정에서 한 가지 결심한 것이 있다면, 우리의 삶에서 출발한 건축의 문법을 찾자는 것이었다. 건축이 삶을 조직하는 공간적 형식이라는 측면에서 스스로의 정의에 비추어볼 때, 삶의 조직이 가장 투명하게 반영되는 지점은 '평면'이라고 생각했다. 평면이라는 것은 그것이 내포하는 형태적, 공간적 문법을 내포하는 것이기에 여기에서 말하고자 하는 평면이란 '평면을 통해 드러나는 건축의 체계'일 것이다. 건축의 체계는 내적인 공간의 질서를 담고 있을 뿐 아니라 그것의 경계면, 즉 도시와의 관계를 포함하고 있다. 서울의 도시한옥, 교토의 마치야, 파리의 아파트 등 전 세계 주택의 모든 유형은 내적인 체계뿐 아니라, 도시에 대한 입장을 드러낸다.
    유형의 탐구, 그 첫 시작은 '채와 간'의 탐구였다. 채와 간의 형식에 주목하게 된 것은 채와 간의 체계로 구성된 한옥의 형식이 이 땅에 오랜 세월 지속해왔을 뿐 아니라, 중국의 사합원 형식을 비롯해 인류의 보편적 건축의 체계로 존재해왔기 때문이다. 채와 간의 체계는 프로젝트를 통해 실험하면서 그 형식은 다양하게 변주되었다. 그리고 한편으로는 채와 간이 갖고 있는 한계를 만났다.

새로운 대안을 찾는 과정에서 직육면체의 '단일매스' 유형에 주목하게 되었다. 단일매스 유형은 좁은 대지, 적은 비용으로도 풍부한 내부공간을 만들 수 있는 장점이 있다. 채와 간의 유형보다 다양한 외부공간을 만드는 것이 불리하지만, 직육면체의 일부를 비우거나 단일매스 주변으로 담을 배치하는 방법을 통해 흥미로운 외부공간을 만들 수 있었다.

'단일매스', 그리고 '채와 간'의 구성을 변형하거나, 때로는 혼합하면서 다양한 공간의 틀을 만들었다. 변형된 평면의 유형들은 곧 다양한 형태와 공간을 낳았다. 그 변형을 낳게 한 것은 우리 시대의 삶의 복합성과 다양성일 것이다. 우리의 삶은 시간에 따라 늘 새로워지며, 새로운 주택의 유형을 생산하고 마침내 유형을 넘어선 '건축'을 요구하게 된다.

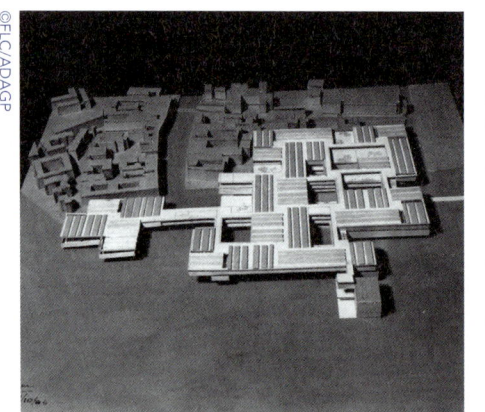 

Fig. 2 (왼쪽) 르 코르뷔지에의 베니스 병원, (오른쪽) 미스 반 데어 로에의 레이크 쇼어 드라이브 아파트
(left) Le Corbusier, Hôpital, Venice, (right) Mies Van de Rohe, Lake Shore Drive apartments

채와 간의 유형

우리 시대 도시주택의 모델을 탐구하고 싶다는 바람은 대학원 시절 줄자를 들고 가회동의 한옥들을 재고 다니던 때부터 키워온 것이다. 우리의 도시한옥이 그랬듯이 '삶과 건축과의 밀접성', 어떤 대지 조건에도 적용되는 '융통성 있는 공간의 문법', 그리고 '정돈된 조형의 원리'를 담아낼 수 있는 도시주택을 제안하고 싶은 소망이 있었다.

1994년, 건축가로 독립하여 시작한 첫 프로젝트가 신도시 일산의 주택이었다. 일산주택이 갖고 있는 평범한 조건을 통해 도시주택에 대한 일반해를 발견하고 싶었다. 그 문법의 기본이 된 것은 '채와 간'이다. 일정한 단위로 만들어진 모듈(간)과 그 모듈이 모여 형성한 하나의 매스(채), 그리고 매스들이 집합해 만든 채와 채가 집합한 건축이다. 간이 모여 채가 되고 채가 모여 전체를 이루는 형식은 동·서양 건축 모두에서 나타나는 건축의 보편적 언어이다. 그렇지만 한옥을 비롯한 동아시아의 목조건축에서 그 체계가 보다 분명히 드러난다. 특별히 한옥이 갖고 있는 채와 간의 체계와 그 미학은 오랜 세월 동안 체화된 익숙한 형식이었기에 중요한 참고가 되었다. 서울의 도시한옥은 다양한 생활을 담을 수 있는 융통성과, 조형적 체계와 구조적 질서를 갖고 있기에 보다 깊이 살펴보았다. 특별히 입구에서 마당으로, 마당에서 대청과 방으로 이어지는 공간의 관계에 주목했다.

동시에 근대건축가들의 작품에서 어떻게 공간을 생성하는지 살펴보았다. 미스 반 데어 로에의 레이크 쇼어 드라이브 아파트에서 기둥과 멀리언의 모듈을 운영하고, 건물을 배치하는 방식이 교훈이 되었다. 특히 두 채의 빌딩을 90도로 틀어 배치한 방식이 이룬 효과가 인상적이었다. 같은 빌딩을 반복해 배치했지만 어느 각도에서 보아도 다른 풍경과 공간감을 선사했다. 르 코르뷔지에의 베니스 병원 프로젝트는 채와 간이 전개되는 방식이 흥미로웠다. 채와 간이 수평적으로 전개되고 수직적으로 적층되는 방식을 배울 수 있었다. 루이스 칸의 킴벨미술관에서는 채와 채 '사이'를 규정하는 방식을 탐구했다. 채와 채 사이에 별도의 영역을 두어, 서번트 스페이스(servant space)로서 기능을 부여하고, 공간적으로도 그에 어울리는 형태를 갖게 했는데 이는 한옥의 전통을 현대적으로 진화시키기를 원했던 나의 고민을 해결해주었다.

유형에 대한 탐색 과정에서 동시대의 건축가들보다는 근대건축가들의 작업을 보다 자세히 살펴보았다. 그들의 작업을 통해 사회와 인간, 도시와 건축의 관계에 대해 깊이 있게 성찰하고, 보편적 체계 안에 특별한 요구를 담아내고자 하는 '건축 유형에 대한 탐구'를 발견할 수 있었다. 또한 동시대 건축가들의 작업에 대해서는 관심을 갖되, 거리를 두는 것이 내 자신의 건축을 찾아가는 데 도움이 될 것이라고 생각했다.

'채'를 설정할 때 먼저 생각하는 것은 몇 개의 채로 주택을 나눌 것이냐이다. 거실과 침실, 식당과 부엌, 서재, 주차장 등 독립된 채로 놓일 영역을 설정하고 그 크기를 가늠해보는 것이 첫 단계다. 채로 공간을 분배하고, 채와 채를 배치하면서 내외부 공간의 관계가 정해진다.

채를 나누면서 동시에 생각하는 것은 채의 크기이다. 채의 크기는 그 안에 배치될 공간의 면적에 의해 정해지지만, '간'(기둥이나 벽체의 간격)의 모듈이 채의 비례를 결정한다. 일산주택의 경우 채의 폭은 3.6m, 규모가 큰 서초동주택은 4.5m를 구사했는데, 간의 모듈이 채의 폭을 만들고 볼륨의 크기를 결정했다. 공간의 크기는 보통 모듈의 단위에 의해서 정해지지만, 경우에 따라서는 채에 붙은 복도의 켜나, 채에서 내민 공간이 덧붙으면서 원하는 공간의 크기를 만들 수 있었다.

채와 채를 잇는 방식 또한 중요한 탐구의 주제였다. 채와 채가 이어지는 사이에, '사이 공간'이 만들어지는데 그 사이 공간은 늘 흥미로운 장소가 되었다. 그 공간은 통로나 작은 여유 공간이 되기도 하고, 때로는 거실과 같은 공간과 이어지면서 넓은 중심 공간의 일부가 되었다.

채를 어떻게 배치하느냐에 따라 다양한 구성이 가능하기 때문에 어떠한 대지의 형상에도 적응할 수 있는 융통성이 있었다. 복층으로 구성된 채는 경우에 따라 두 층 이상으로 열린 공간을 만들 수 있기 때문에 필요한 경우 높은 볼륨의 공간을 만들 수 있었다. 채와 간의 구성을 통해 만들어내는 공간은 건축역사 내내 있었던 보편적 문법이지만, 필지의 형태와 프로그램의 내용, 환경에 대응하는 방식을 통해 얼마든지 지역성을 표현할 수 있다고 믿었다. 보편적인 건축언어를 구사하면서도 지역성을 담아내고, 한편으로는 개인과 집단의 요구를 공간에 담아내는 문법이라는 점이 채와 간의 체계가 갖는 큰 장점이었다.

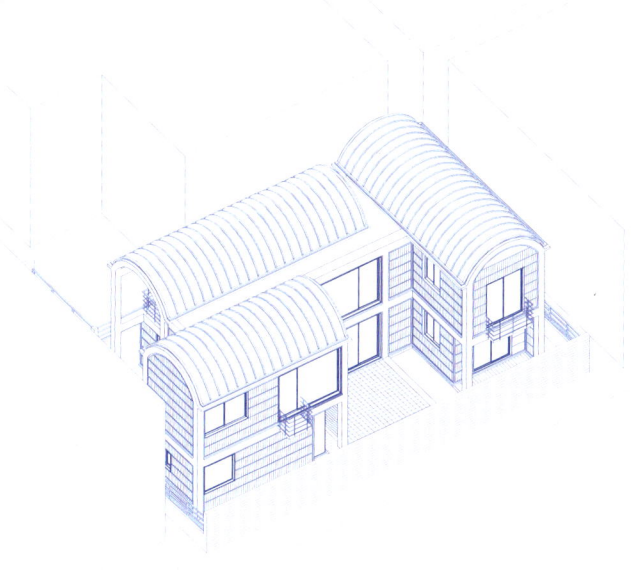

Fig. 3 일산주택의 경우 3.6m로 간의 모듈이 채의 폭을 만들고 볼륨의 크기를 결정했다.
The width of Chae at Ilsan Residence these are 3.6m. Modeul of the Kan creates the width of Chae and size of volume.

The Exploration of Types

일산주택 : 공간의 체계와 경험의 서사

채와 간에 대한 탐구를 가장 잘 보여주는 작업은 일산주택이다. 일산주택은 1994년 사무실을 개소하고, 처음으로 수주한 프로젝트이다. 설계 시작부터 현대건축의 형식으로 만든 이 시대의 '한옥'을 목표했으며, 대학생 때부터 품어온 오랜 소망을 첫 작품을 통해서 이루고자 했다. 생각의 한편에는 우리의 전통한옥이 있었고 또 한편에는 근대건축이 성취한 유산이 있었다. 그리고 물론 이 시대 가족의 삶의 방식에 주목했다.

먼저, 3.6m 폭을 가진 볼트형의 균질한 볼륨 세 개를 설정하고 그 속에 필요한 공간들을 조직하면서 계획안을 만들어 나갔다. 세 채의 덩어리(mass)는 놓이는 방식에 따라 어떠한 대지에도 적응할 수 있는 집으로 만들 수 있었다. 거실과 침실 등의 세부 영역의 분화와 연결에도 잘 맞았고, 외부와 내부 공간의 다양한 관계를 만들어가는 데도 유리했다. 하나의 채 안에 다양한 높이와 단면을 가진 한옥의 풍부한 공간감을 현대의 언어로 진화시키고 싶었다. 볼트 천장이 노출된 2층 높이의 거실, 거실의 볼륨과 대비되는 낮고 긴 복도, 1층 볼트 높이의 가족실, 안방과 식당 앞의 75cm 폭의 툇마루, 정발산으로 트여 있는 볼트 아래의 발코니, 거실과 연결된 아늑한 안마당 등 다채로운 공간을 제안했다. 세 채로 구성된 주택의 공간은 3대가 함께 사는 가족들의 독립된 영역을 구축하면서도, 거실과 마당 등 공동의 공간을 형성하기에도 유리했다.

단순한 세 개의 덩어리들은 내부 공간뿐 아니라 외부, 가로와 도시를 향해 나름의 표정을 지었다. 같은 크기의 덩어리지만 축의 방향을 돌리거나 조금씩 어긋나게 하여, 어느 방향에서도 다른 모습을 연출할 수 있었다. 2층 높이로 열린 진입부, 돌출된 발코니와 처마, 그리고 필로티 공간이 도시와 접촉면을 만들고, 다양한 도시적 관계를 표현하는 입면을 구성했다.

채와 간이 구조와 공간을 구축하는 원리가 되었고, 채와 간의 조합은 다양한 외부 공간을 만들어내었다. 마당과 내부 공간의 관계 속에서 다양한 스케일과 공간감으로 전개되었다. 일산주택은 진입마당, 중정, 그리고 부엌쪽 서비스 마당,

Fig. 4 세 채로 구성된 주택의 공간은 3대가 함께 사는 가족들의 독립된 영역을 구축하면서도 거실과 마당 등 공동의 공간을 형성하기에도 유리했다.
The three separate Chae provided private spaces for the three generations to live with one another while at the same time also providing shared spaces in the living room and the court.

Fig. 5 일산주택은 도시에서 길로, 길에서 집의 현관으로, 현관에서 거실로, 거실에서 방으로 이어지는 서사 구조에 집중했다.
I focused on the narrative structure of moving from the city to each road, each road to the entrance hall of the house, from the entrance hall to the living room, from the living room to the bedroom.

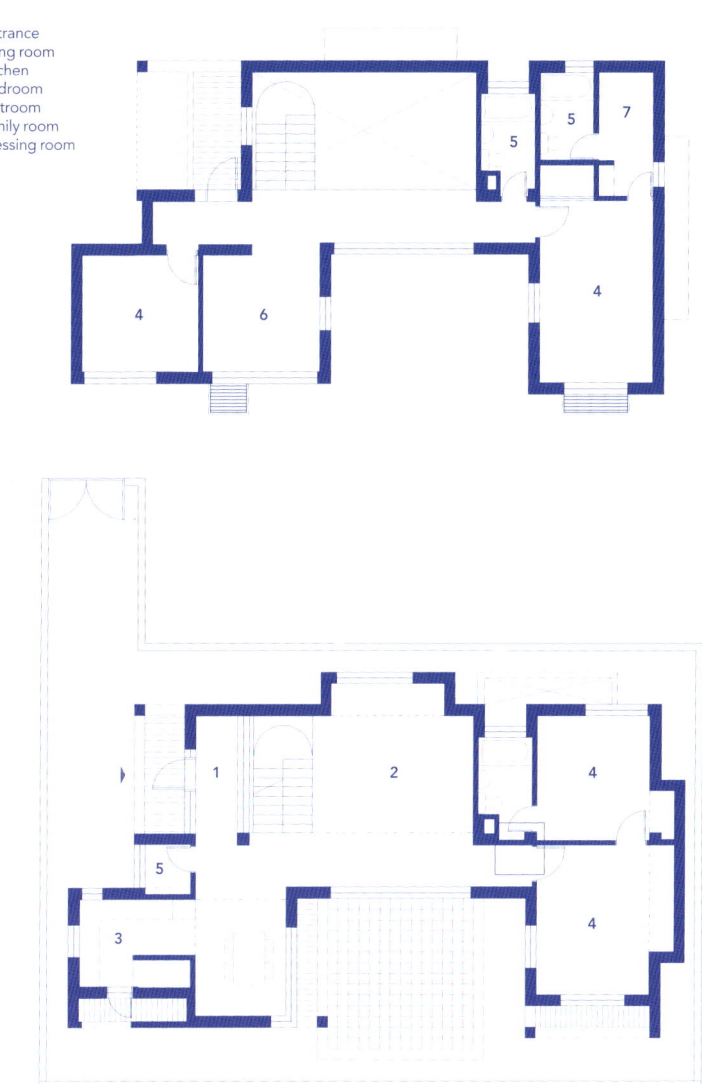

Fig. 6 일산주택 평면도 / 모든 공간은 외부1-내부-외부2로 이르는 관계망 속에 자리 잡는다.
Ilsan Residence plan / All spaces are placed within the relationship of the public and the private.

안방 마당, 크게 네 종류의 마당이 있다. 그 마당과 내부공간이 만나는 방식은 각각 달라서 작은 집이지만 풍부한 공간을 경험할 수 있다. 모든 내부 공간은 나름의 외부를 갖는다. 현관은 진입마당, 거실과 식당은 툇마루와 중정, 안방은 툇마루와 안방마당, 부엌과 다용도실은 서비스 마당, 2층 방들도 발코니와 테라스를 갖는다. 모든 공간은 외부1(public)-내부-외부2(private)로 이르는 관계망 속에 자리 잡는다.

일산주택은 도시에서 길로, 길에서 집의 현관으로, 현관에서 거실로, 거실에서 방으로 이어지는 서사 구조에 집중했다. 2개층 높이로 열린 입구에서 1층 높이의 현관이 이어지고, 천장이 낮은 현관을 지나면 다시 2층 높이로 열린 거실을 만난다. 다시 1층 높이의 복도를 지나 방문 앞에 전실을 지나 방으로 이어진다. 공간이 전개되면서 도시의 공간 스케일에 대응하거나, 내부 공간의 격식에 따라 공간의 크기와 빛의 강약이 조절된다. 주공간과 진입 공간, 그리고 영역 사이의 결절부로 구성한 공간의 요소는 동·서양의 고전건축에서 배운 것이다. 대지 입구에서 거실, 2층 가족실에 이르는 여러 마디의 공간의 전개는 일산주택의 채와 간이라는 객관적 체계 속에서 주관적으로 경험하는 공간의 내러티브가 가능하다는 것을 증명한다. 근대건축에서 진화한 중성적인 구축체계에 고전적인 서사를 담을 수 있는 것이다. 나에게 건축은 구축체계와 공간 서사의 결합이다. 가장 단순한 도시주택이라 하더라도 공간의 체계와 경험의 서사라는 두 개의 차원을 지닌다. 전형 또는 유형이 될 수 있는 건축은 단순한 얼개 속에 풍부한 내용을 담을 수 있어야 한다.

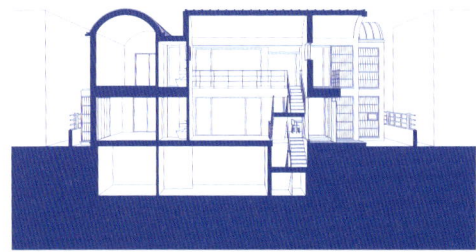

Fig. 7 공간이 전개되면서 도시의 공간 스케일에 대응하거나, 내부 공간의 격식에 따라 공간의 크기와 빛의 강약이 조절된다.

As the space develops, the size of the space and the amount of light accedes to and is adapted to the spatial scale of the city and the interior logic of the space.

Fig. 8 (위) 돌체하우스, (아래) 서초동주택 / 일산주택 이후 돌체하우스와 서초동주택 등의 프로젝트에 비슷한 아이디어를 전개했다.
(top) Dolce House, (bottom) Seocho Residence / I developed similar ideas projects like the Dolce House and the Seocho Residence.

일산주택의 체계, 그 전개와 진화

1994년 일산주택의 계획안을 만들고 나서, 하나의 보편적 체계로서 채와 간의 유형을 다른 프로젝트에 충분히 응용할 가능성이 있다고 보았다. 이후, 돌체하우스, 서초동주택 등의 프로젝트에 비슷한 아이디어가 전개되었다. 채의 크기에 변화를 주면서 원하는 규모의 공간을 확보할 수 있었고, 아담한 마당을 채와 채 사이에서 만들 수 있었다. 일산주택의 체계는 주택이 아닌 일반 건축에도 충분히 적용할 수 있다고 생각했다. 1995년 보건복지부에서 보건소와 보건지소에 대한 표준설계현상이 있었는데, 일산주택과 같은 체계를 적용한 안으로 제출하여 당선되었다. 일산주택에서 확립한 유형은 어떠한 대지에도 적용할 융통성이 있었고, 구조적, 조형적 체계를 갖고 있었다. 이후 2년 동안 지역보건소, 보건지소에 일산주택에 적용했던 시스템을 적용하여 여러 프로젝트를 완성했다. 철원군 보건소, 경주시 보건소, 영동군 학산지소 등이 대표적인 예이다.

간을 형성한 모듈을 반복하여 채를 만들고, 채의 집합이 전체 건축이 되는 방식은 2002년 이우학교 설계에서 새롭게 발전되었다. 이우학교는 8.4m×1.8m 모듈의 간으로 구성된 여러 채의 건물로 이루어졌다. 교사동을 여러개로 나눈 배치는 섬세한 지형에 적응하는 데 유리했다. 교사동 사이에 저절로 생겨난 마당은 학교의 특징이 되었다. 이우학교에 구사된 철골 구조와 건식 벽체는 공장제작의 프리패브리케이션 방식으로 만들어졌다. 채와 간의 형식이 현대적인 생산방식을 통해 빠른 속도로 제작될 수 있음을 확인했다.

Fig. 9 1995년 공공보건의료기관 표준설계 당선안. 일산주택의 체계를 적용한 프로젝트
In 1995, the Ministry of Health and Welfare standard design

Fig. 10 보건소와 보건지소의 표준설계도
Health Care Centers and Hospitals standard design plan

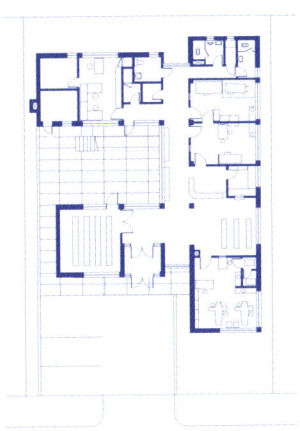

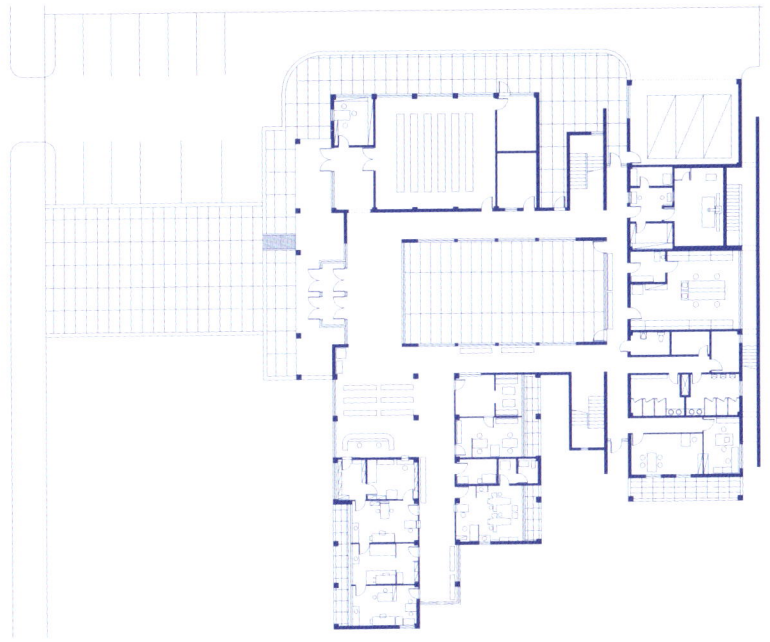

The Exploration of Types

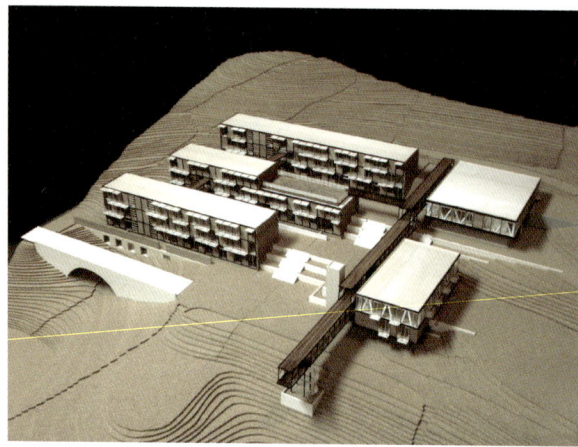

Fig. 11 이우학교는 8.4m × 1.8m 모듈의 간으로 구성된 여러 채의 건물로 이루어졌으며, 현대적인 생산방식을 통해 빠른 속도로 제작할 수 있었다.

The Ewoo School is composed of modular Kan measuring 8.4m × 1.8m. The placement of each separated school annex made it easy to adapt to the shape.

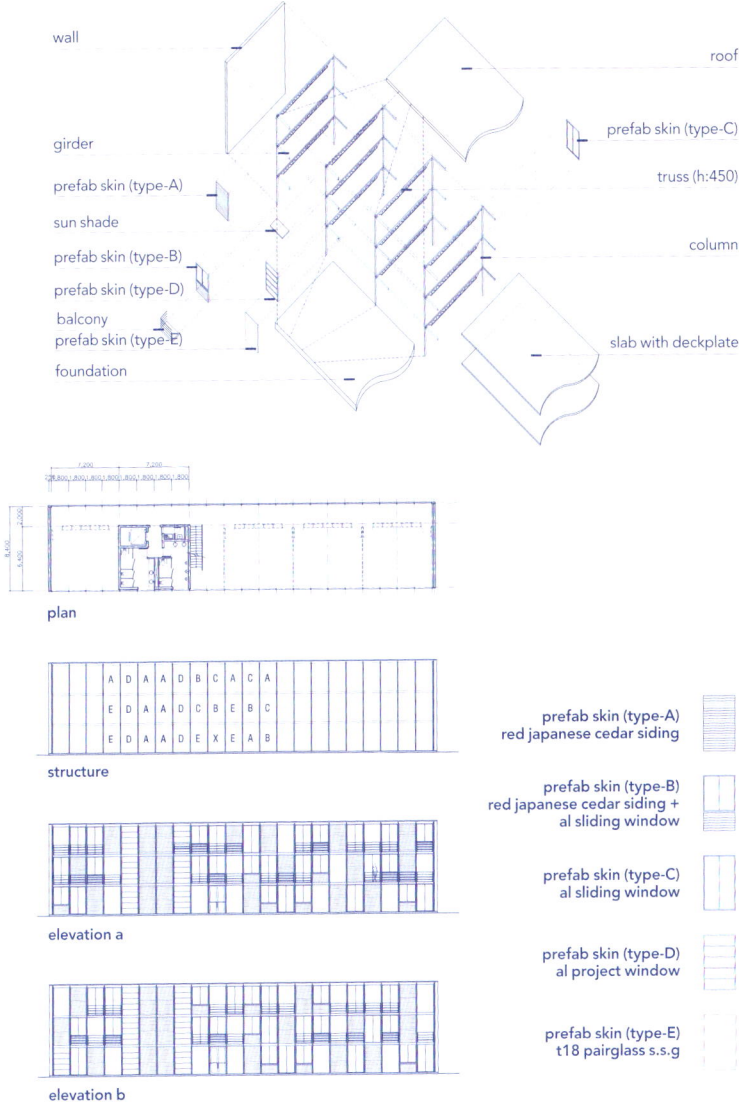

Fig. 12 이우학교 평면도와 철골구조도
Ewoo School plan and steel frame plan

단일매스-박스 유형

일산주택에 적용되었던, 채와 간의 유형이 보건소와 학교 등 다른 영역으로 확장되었다면 '단일매스-박스' 유형은 보건소, 업무시설 등에서 시작해서, 주택에 적용하게 된 경우이다. 채와 간, 단일매스 유형은 건축을 생성하는 보편적 체계이기에, 교회, 의료시설, 학교 등 다른 영역과 서로 넘나들면서 진화했다. '채와 간'이 한국의 전통주택 한옥에 대한 탐구에서 시작되었다면, '단일매스-박스'는 현대적인 '일반건축물'에서 시작되었다.

채와 간 유형에 대한 제안을 주택을 거쳐 보건소, 학교 프로젝트에 이르기까지 다양하게 전개하면서, 몇 가지 한계를 느꼈다. 채가 전개되면서 다양한 마당을 만들어내기는 쉽지만, 그 과정에서 외피의 면적이 커지고 디테일이 복잡해지면서 공사비가 늘어났다. 보건소와 의료원 등 공공건축에서는 공사예산을 맞추는 것이 중요한데 채와 간의 체계로는 공사비를 맞추기가 어려웠다. 채와 간의 형식은 홑집(single layered space)의 형태로 공간의 두께가 상대적으로 얇고, 따라서 내부 공간의 풍성함을 만드는 데 불리했다. 구조적으로는 한 베이(single-bay) 단순보가 많이 사용되면서 기둥과 보의 크기를 줄이는 데에도 한계가 있었다. 또한 채에서 채로 이동하는 통로 공간이 많이 필요했기에 면적 배분에서도 어려운 점이 많았다.

채와 간의 체계와 다른 대안을 찾아야만 했다. 외피 면적을 줄이고 통로 공간을 최소화할 수 있는 효율적인 공간 형식이 필요했다. 여러 차례의 시행착오 끝에 찾은 형식은 직육면체 단일매스였다. 단순한 큐브 형태는 표면적이 작기 때문에 경제적이다. 통로 등 공용 공간을 효율적으로 만들 수 있어서 같은 면적으로도 훨씬 여유로운 공간이 생겨났다. 또한 볼륨이 두터워지면서 내부 공간에 다양한 공간을 연출할 수 있는 가능성이 열렸다.

전통건축에서 구사된 채와 간의 형식은 긴 스팬을 만들기 어려웠던 전통적인

Fig. 13 과천주택은 직육면체의 단순한 형식으로 설계를 시작했다 하더라도, 마당과 테라스 발코니 등을 만들면서 복잡한 형태로 마무리되었다.
Although design began as a simple cube, the shape became more complicated with the addition of a courtyard and balcony.

The Exploration of Types

구조체계와 관계가 있다. 근대건축이 만들어낸 철골 구조와 철근콘크리트 구조가 선사하는 큰 볼륨은 채와 간의 집합을 하나의 입체 안에 넣을 수 있게 해주었다. 하나의 커다란 볼륨이 만들어졌기 때문에 그 안에서 레벨의 변화와 공간의 강약을 다채롭게 만들 수 있었다. 단일매스-박스 유형을 건축의 문법으로 탐구하게 된 것은 이렇듯 절박한 상황 속에서 필연적으로 찾은 결과였다.

단일매스-박스 유형의 시작 : 고성군보건소

'단일매스-박스' 유형을 의식하고 최초로 한 작업은 1998년 설계한 고성군보건소이다. 1995년 보건소 표준설계에 당선된 이후, 여러 종류의 공공의료기관을 설계하면서 공사비를 맞추는 일로 여러 해 고통을 겪은 뒤였다. 고성군의 예산이 극도로 제한적이었으므로 설계 시작부터 단순한 직육면체를 염두에 두었다.

    단순한 육면체 공간을 두 층으로 나누어 공간을 배치하고, 중심 공간은 크게 열어 한눈에 전체 공간이 보이도록 계획했다. 결과적으로는 예산도 절감하고 면적을 효율적으로 사용하면서도, 충분한 크기의 공용 공간, 풍부한 공간감을 만들 수 있었다. 이러한 형식은 문경시 보건소 등의 프로젝트로 이어지면서 단일매스, 단순한 공간이 선사하는 매력을 탐구할 수 있었다.

    고성군보건소의 구성 논리는 근대건축에서 진화해온 프리플랜(free-plan)의 원리를 담고 있다. 기둥이 외벽과 독립적으로 서 있기 때문에 외벽과 내벽은 하중을 받는 임무를 면제받아 어떤 위치에 어떤 형태로도 설 수 있다. 그렇지만 프리플랜의 원리를 모든 공간에 적용할 의도는 없었다. 분할되는 작은 공간에서는 기둥을 벽과 맞추어 내부 공간의 활용도를 높였고, 크게 열린 중심 공간에서는 기둥을 드러내어 공간의 질서를 보여주었다.

    단일매스 유형을 통해 추구하는 것은 프리플랜이 결코 아니었다. 그 시작은 외벽 면적을 줄여 공사비를 절약하고, 깊은 공간을 만들어 공간적 가능성을 확장하고, 공용 공간을 효율적으로 만드는 것에서 출발했기 때문이다.

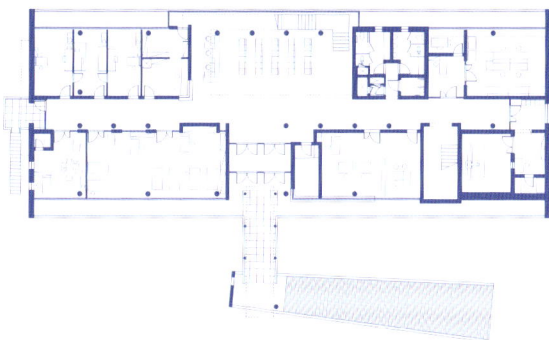

Fig. 14 고성군보건소 평면도 / 고성군보건소의 구성 논리는 근대건축에서 진화해온 프리플랜의 원리를 담고 있다. 기둥이 외벽과 독립적으로 서 있기 때문에 외벽과 내벽은 하중을 받는 임무를 면제받아 어떤 위치에 어떤 형태로도 설 수 있다.

Health Care Center Goseong plan / The composition system of the Health Care Center Goseong contains principles that have evolved from the free-plan of modern architecture. As the columns stand independently of the exterior walls, the interior walls and the inner walls, both interior and exterior walls are excused from their load-bearing role and can be placed in any form.

단일매스-박스 유형의 주택

공공건축에서 단일매스 형식은 기둥 간격이 기준이 되는 모듈의 체계로 만들어졌다면, 주택의 경우는 모듈의 전개가 처음부터 불가능했다. 주택의 규모가 작기 때문에 모듈의 반복이 이루어지기 어려웠고, 방과 거실, 욕실과 주방 등 주택의 개별 공간을 일정한 모듈로 묶는 일은 불가능했다. 외벽면이 주택의 구조를 감당하고 그 내부에 꼭 필요한 곳에 구조가 서는 방식이 구사되었다. 미스 반 데어 로에의 투겐타트 주택이나 르 코르뷔지에의 빌라 사보아의 경우에서 사용된 방식- 격자 그리드를 따라 기둥이 배치되고, 벽이 자유롭게 형성되는 형식과는 크게 다른 방식이다.

근대건축의 사례 중에서는 아돌프 로스의 뮐러주택의 형식과 보다 더 유사하지만, 공간 구성에는 차이가 있다. 뮐러주택은 박스의 내부에 다양한 높이의 단면이 적층되면서 공간의 변화를 만들고, 고유한 성격을 지닌 각 공간들이 계단을 매

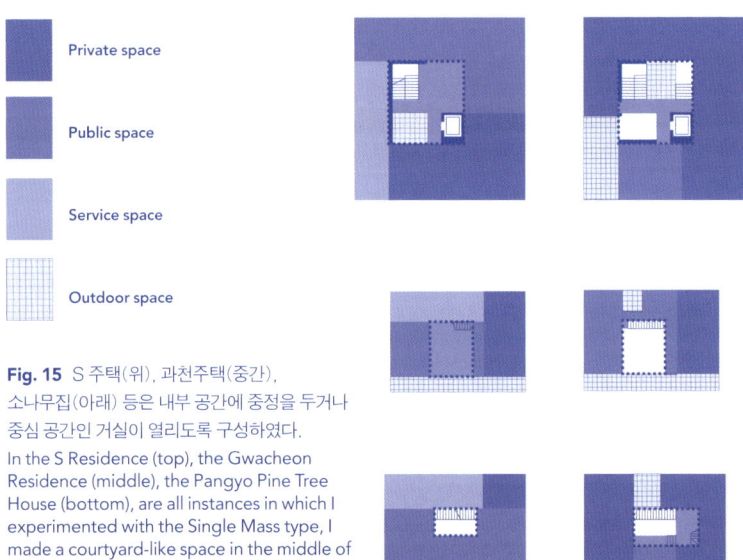

Fig. 15 S 주택(위), 과천주택(중간), 소나무집(아래) 등은 내부 공간에 중정을 두거나 중심 공간인 거실이 열리도록 구성하였다.
In the S Residence (top), the Gwacheon Residence (middle), the Pangyo Pine Tree House (bottom), are all instances in which I experimented with the Single Mass type, I made a courtyard-like space in the middle of the house or opened up the focal living room to create a courtyard-like space.

개로 연결된다. 과천주택, 성북동주택, 소나무집 등 단일매스 유형을 실험한 주택의 경우, 내부 공간에 중정을 두거나 중심 공간인 거실이 열리도록 구성하여 결국에는 중정의 성격을 지닌 공간이 중심에 만들어진다. 한옥에서 진화한 중정형 도시주택의 공간 구조가 직육면체 볼륨 안에 담긴다는 점에서 근대건축의 사례와는 크게 구별된다. 또한 침실들이 같은 영역에 모여 있는 서양의 주택과 달리, 부모의 침실과 자녀 침실은 멀리 떨어지고, 그 사이를 거실이나 가족실이 채운다. 이러한 배치는 가족 커뮤니티 중심을 원하는 동시에 가족 간에 심리적, 물리적 거리를 유지했던 전통이 반영된 것이다.

직육면체의 단순한 형식으로 설계를 시작했다 하더라도, 마당과 테라스 발코니 등을 만들면서 복잡한 형태로 마무리되었다. 사람과 사람, 도시와 자연 간에 다양한 관계를 맺어야 하는 주택은 마지막에 섬세한 체계를 지닐 수밖에 없다.

Fig. 16 과천주택은 제한된 크기의 대지에 경제적으로 주택을 만들어야 하는 숙제를 해결하기 위해서 단일매스 형식이 요구되었다.
The Gwacheon Residence was requested as a solution to satisfy the requirements of building on a limited area of land, as economically as possible.

과천주택은 제한된 크기의 대지에 경제적으로 주택을 만들어야 하는 숙제를 해결하기 위해서 단일매스 형식이 요구되었다. 두 개층 높이의 단순한 직육면체의 볼륨 위에 경사지붕이 올라간 구성이다. 외벽면을 얇은 두께의 구조벽으로 만들고 1층 계단 뒷벽에 내부 기둥을 두어 전체적인 구조의 틀을 조직했다. 구조 기둥의 단면을 얇게 구사하여 벽두께 안에 넣었기 때문에 기둥은 내부 공간에서 드러나지 않는다.

직육면체의 공간이 두 층으로 나뉘고 거실 부분이 두 개층 높이로 크게 열려 있다. 과천주택의 거실은 내부화된 마당으로 다양한 행위의 중심이 된다. 열린 거실의 둘레로 식당과 부엌, 안방과 서재, 자녀방 등이 배치되는데, 거실이 마당이라고 가정한다면, 그 구성은 마치 도시한옥의 구성과 닮았다고 할 수 있다. 직육면체에 안에 주택을 담는 경우, 기둥의 질서는 벽 안으로 숨기 때문에 간의 질서가 거의 보이지 않는다. 다만 거실이나 중정을 기준으로 공간이 분절되면서 3X3 또는 3X2의 공간의 질서가 생성된다.

Fig. 17 과천주택 평면도 / 열린 거실의 둘레로 식당과 부엌, 안방과 서재, 자녀방 등이 배치되는데, 거실이 마당이라고 가정한다면, 그 구성은 마치 도시한옥의 구성과 닮았다고 할 수 있다.
Gwacheon Residence plan / Arranged around the perimeter of the living room are the dining room and kitchen, master bedroom and study, and the children's rooms, just as if the living room were the courtyard of a urban traditional *hanok*.

Fig. 18 거실과 안방, 식당으로 이어지는 공간은 하나로 열려 대공간을 누릴 수 있을 뿐 아니라, 미래에도 공간의 구획을 쉽게 조정할 수 있는 가능성을 갖는다.
The space connecting the living room, the master bedroom and the dining room converges into a large space, allowing space partitioning to be easily adjusted in the future.

S 주택은 간결한 육면체의 윤곽 안에 다양한 굴곡을 담고 있으며, 3대가 함께 사는 가족을 위한 공간으로 구성되었다. 각 세대가 고유한 영역을 가지면서도 서로 교류하는 방식, 단일매스 표면을 넘어 외부 공간과 내부 공간이 소통하는 방식, 단일매스 내부의 서로 다른 레벨과 볼륨이 만들어내는 풍요로운 변화, 단일매스 박스 내부로 끌어들인 네 개의 다른 마당 등, S 주택은 풍부한 공간의 레퍼토리를 품고 있다. 직육면체, 단순한 윤곽을 가졌기에 오히려 공간의 자유로운 변화를 담아낼 수 있었다.

건축주는 가족구성원의 변화에 따라 공간 구조를 쉽게 고칠 수 있는 집을 원했는데, 이는 집의 공간 체계와 구조 시스템을 통해 구현되었다. 집 전체 윤곽을 형성하는 외곽 사각형과 집 내부에 또 다른 작은 사각형으로 구조가 짜인다. 두 사각형 사이는 미래의 변화를 수용할 수 있는 자유로운 공간이 된다. 내부의 작은 사각형 안에는 계단과 엘리베이터와 더불어 중정이 들어있다. 작은 공간이지만 여

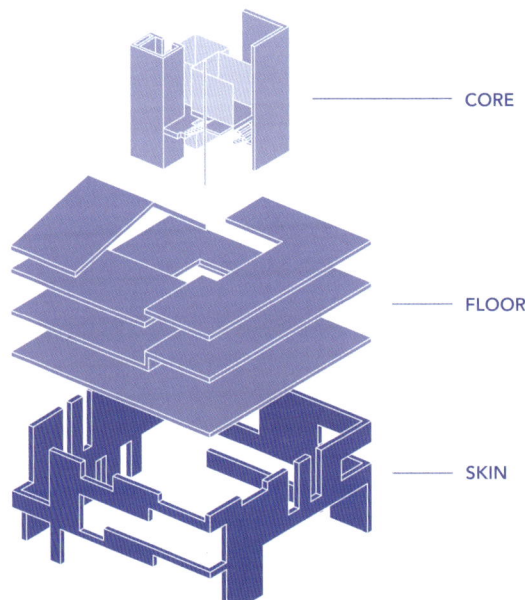

Fig. 19 S 주택은 단순한 윤곽을 가졌기에 오히려 공간의 자유로운 변화를 담아낼 수 있었다.
The very simplicity of the lines of the S Residence enables the spatial flexibility and freedom there within.

러 공간의 다발이 묶여 있는 곳이다. 계단을 오르내리거나 중심 공간을 가로지르면서 다양한 공간을 경험하고 풍부한 건축적인 장소를 누린다. 내부의 작은 사각형은 ㅁ자 도시주택의 중정의 역할을 수행하며 그 둘레로 집의 각 공간이 전개된다. ㅁ자 주변으로 거실과 주방, 입구, 주방과 서비스, 안방 등으로 구별된 영역이 배치되며, 각각의 영역은 마치 독립된 채처럼 분명히 구별되지만, 하나의 사각형 윤곽 안에 가지런히 놓여 있다.

Fig. 20 S 주택 평면도 / ㅁ자 주변으로 거실과 주방, 입구, 주방과 서비스 안방 등으로 구별된 영역을 배치했다.

S Residence plan / Around the square of this central space, the demarcated spaces of the living room and dining room, the entrance hallway, the kitchen and service room, and the master bedroom are arranged.

1. entrance
2. living room
3. kitchen
4. bedroom
5. workroom
6. restroom
7. courtyard
8. family room
9. dressing room
10. study

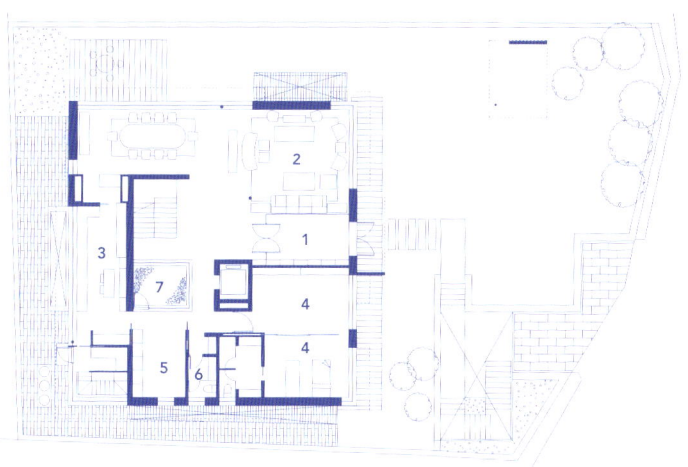

소나무집은 가장 단순한 박스에서 출발하여 건축주의 요구와 대지와 그 주변의 상황에 부합하는 원리들이 차곡차곡 쌓인 곳이다. 단순한 박스이지만 삶의 모습은 다양하고 풍부하므로 자연스럽게 박스의 모습은 새롭게 진화되었다. 거실, 부엌, 침실, 다락 등, 원하는 높이에 따라 다양한 레벨이 생겨났다. 분절된 레벨은 입면에 투영되어 다양한 높이의 창문으로 표현되며 박스의 형태는 미묘하게 변화한다. 성격이 다른 마당이 박스 안에 삽입되면서 박스의 형상은 더욱 섬세하게 변형되어 재미있는 형태로 진화되어갔다. 전망을 볼 수 있는 2층 발코니, 2층으로 향한 직선계단이 열려 보이는 볼륨이 큰 보이드, 부부만을 위한 중정, 다락과 이어지는 옥상의 마당 등 여러 공간이 박스 속에 담겨 있다.

Fig. 21 박스 안에 여러 공간과 기능이 중첩되면서 소나무집은 서로 다른 세계의 공존, 서로 다른 존재가 함께 있음을 드러냈다.
As the spaces within the box are used for multiple functions, the Pangyo Pine Tree House becomes a place where different worlds co-exist, where it becomes apparent that different existences are together.

박스 안에 여러 공간과 기능이 중첩되면서 집은 서로 다른 세계의 공존, 서로 다른 존재가 '함께 있음'을 드러냈다. 서로 다른 공간들은 낭비되는 공간을 최소화한 연결 방식으로 묶여져 있어 효율적인 평면을 구성한다. 서로 다른 것을 '다른 그대로' 담아내는 형식, 그리고 그것을 공간의 낭비 없이 효율적으로 엮어내려 한 의도가 반영되었다.

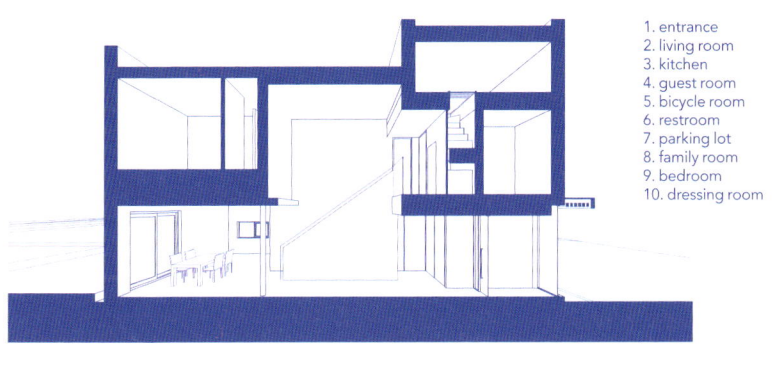

1. entrance
2. living room
3. kitchen
4. guest room
5. bicycle room
6. restroom
7. parking lot
8. family room
9. bedroom
10. dressing room

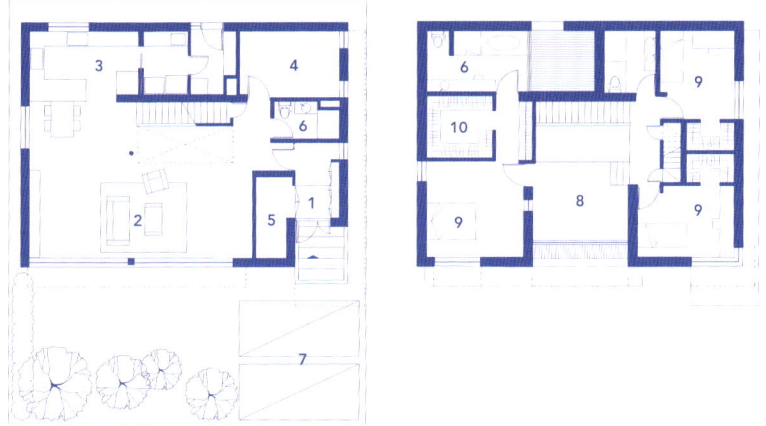

Fig. 22 (위) 소나무집 단면도, (아래) 소나무집 평면도 / 거실, 부엌, 침실, 다락 등 원하는 높이에 따라 다양한 레벨이 생겨났다.
(top) Pangyo Pine Tree Residence section, (bottom) Pangyo Pine Tree Residence plan / The living room, kitchen, bedroom, attic and so on have different ceiling heights which produce a variety of levels.

Fig. 23 분절된 레벨은 입면에 투영되어 다양한 높이의 창문으로 표현되며 박스의 형태는 미묘하게 변화한다.

Separated levels are projected onto the façade and create various expressions of window treatments which subtly change the appearance of the cube.

유형 - 보편적 체계와 특별한 건축

건축가 자신의 알리바이로만 존재하는 담론은 생명력을 갖지 못한다. 시대와 지역이 공유하는 보편성에 대한 성찰이 담겨 있을 때, 공감을 얻고, 비판을 받고, 진화할 수 있다. 건축은 건축가의 '작품', 개인의 '재산'을 넘어서 공동체의 자산으로 존재한다. 그렇기에 공동체의 소통을 가능하게 하는 기반을 건축 안에서 찾는 일이 중요하다.

유형에 대한 탐구는 건축 안에 있는 보편적 논의의 기반을 찾아가는 하나의 방법론이자 실천이다. '채와 간', 그리고 '단일매스-박스'라는 이름으로 제시한 유형은 건축을 생성하는 원리를 포함한다. 그 안에는 공간의 관계, 구조의 체계와 조형 운영의 원리가 담겨 있다. 여기서 '유형'이란 '공간을 만드는 문법'이라고 정의할 수 있다. 그 문법에는 오랜 세월 건축의 역사 속에서 이어져온 보편성과 융통성, 그리고 어떤 체계가 필요하다. 그리고 건축가의 프로젝트를 통해 새롭게 해석되고 진화하는 가능성이 열려 있어야 한다.

서구에서 시작된 근대건축과 현대건축을 대면하면서, 그들 건축의 문장과 언어를 만들어낸 문법을 그대로 차용하기보다는 그것에 대해 이해하고, 우리의 삶의 조건에 비추어 질문하고, 발전시켜야 한다고 믿었다. 새로운 클라이언트, 새로운 대지는 언제나 특별한 해결을 요구하지만, 한발 들어가면 보편적인 가치와 공유하는 조건을 내포한다. 그 보편성에 기대어 얼마든지 특별한 해결안을 낼 수 있다. 요리는 특별하지만, 그것을 담는 그릇이나 코스의 구성은 얼마든지 보편적일 수 있다. 유형에 관한 탐구가 건축을 하나의 정답으로 정리하려는 노력으로 보일지 모르지만, 외려 그 반대이다. 건축을 생성하는 문법에 대한 탐구를 통해 우리 시대, 우리 지역의 특수성과 정체성을 찾아가는 징검다리가 놓일 수 있다. 그 징검다리를 통해 특별한 취향과 비전이 담긴 고유한 건축이 만들어질 수 있다.

Chae and *Kan* and the Single Unit Mass – Box

Throughout my architectural training, several different architectural perspectives and theories were hotly debated. Modernism, Structuralism, Contextualism, Typology, Postmodernism, Late Modernism, Deconstructivism, Dutch Modern, Neo-urbanism and so on were all interesting and worthy theories, as was the attendant architecture, but none of these felt completely right for me. Instead of appealing to me as a unified world they felt like fragments, elements that seemed to congregate or pass me by like the wind. If the most important factor for the formation of an architect's own voice is a point of view about the world and architecture, I believed that this perspective could only be built on the fundamentals of reality after time had been spent in confronting the realities of life.

If there is one thing that I resolved during this process, it was to discover the discipline and process of an architecture that is inspired by our lives. If architecture can be defined as spatial logic with which one can order one's life, this is made most apparent by the floor plan. The floor plan, containing both the form and spatial layout of the building, here becomes the plan through which the system of architecture becomes evident. The system of architecture not only contains the logic of the inner space but also the demarcations, in short its connections to the city. The urban *hanoks* of Seoul, the Machiya of Kyoto, the apartments of Paris—all these residential spaces not only have spatial logic but also manifest a point of view about the city in which they are located.

The beginning of my exploration into form is Chae (each wing or building in a traditional Korean home) and Kan (the term used to describe the intervals within each of those wings or buildings in a traditional Korean home). My reasons for focusing on Chae and Kan were not only because the form of the *hanok* is many centuries old but also because, as evidenced by the Chinese Siheyuan, it has existed as a universal architectural system. Experimenting with the system of Chae and Kan through projects, I was able to transform that form in a variety of new ways. I also discovered the limitations of Chae and Kan.

In the process of looking for a new solution, I came to focus on a single mass form. The benefit of a single mass form is that it can provide a plentiful interior space on a small piece of land and for a limited budget. Although it has less external diversity than the concept of Chae and Kan, it was still possible to make an interesting external space by emptying a part of the cubed shape or by placing a wall within the vicinity of the single mass form.

Using adaptations of the single mass form or the structure of Chae and Kan, and at times using a combination of the two, I created a range of spatial frameworks. Adaptations of these shapes and surfaces soon gave birth to a variety of shapes and spaces. The reason for those changes and adaptations is the complexity and variety of our modern lives. Our lives change with the times, and demand new types of houses and finally, go beyond that to demand a new kind of architecture.

The Type of *Chae* and *Kan*

My desire to explore the models of modern urban residential housing is one I have nurtured since graduate school, when measuring the *hanok* buildings of Gahoe-dong in Byukchon. (Fig. 1) My hope is to propose an urban house that will show the congruence of life and architecture, a flexibility of spatial mechanics that can be adapted to any site, and has the principles of orderly shapes, just as our urban traditional *hanok*s do.

In 1994, I undertook the first project of my new fledgling practice, a house in the new city of Ilsan. In the typically ordinary conditions of Ilsan Residence I hoped to discover a general solution for urban housing. The fundamental discipline and process of that solution was the form of 'Chae' and 'Kan'. I used regular units of modules (Kan) and groupings of those modules as a mass (Chae), and grouped those masses to complete the construction. Grouping units of Kan to create Chae, and grouping Chae to create the whole is a universal vocabulary of architecture in both eastern and western architecture. However, this system is particularly evident, in East Asian wooden architecture, including the Korean *Hanok*. The *Hanok* was a particularly important point of reference for me as it was an aesthetic that I both understood and with which I felt familiar. I observed that the urban *hanok*s of Seoul combined flexibility with a system of shapes and structural logic. I paid special attention to the space between the entrance to the inner court, from the inner court to the main room and adjoining rooms.

At the same time, I also observed how contemporary architects created spaces. I studied the modules created from the columns and mullions in Mies Van de Rohe's Lake Shore Drive apartments, and its building's placement. (Fig. 2) In particular, the idea of placing two buildings at a sharp right angle made an impression on me. Although identical buildings were simply repeated and placed, the different angle of the building created new and different aspects and sense of space. Le Corbusier's Hôspital in Venice project also has an interesting development of Chae and Kan. (Fig. 2) In this instance, I learned the way in which Chae and Kan could be developed horizontally and stacked in a straight line. I explored the interstitial spaces between Kan in Louis Kahn's Kimbell Art Museum. By placing spaces between each Chae, it utilized them as a servant space, giving them a space to suit the shape, and these aspects showed me how the modern evolution of a traditional *hanok* might look.

During my exploration of architectural types, instead of looking at my peers I looked in great detail at Modern architects. This was because in their work I found insight into society and man, the city and architecture, and an exploration of architectural shapes in a universal architectural system. Also, although the work of my contemporaries interested me, I felt that an appropriate distance was helpful in discovering my own architectural style.

When deciding on Chae, the first thing to consider was how many Chae were necessary for the building. Deciding on the placement and size of each independent Chae for the living room and bedroom, dining room and kitchen, study, and parking space is the first step. Dividing the space into Chae and then the placement of each Chae decides the relationship between the interior of the home with the outside.

A simultaneous consideration with the division of Chae is the size of each Chae. The size of each Chae is affected by the size of the space within which it will be placed: the Kan (the spacing of columns and walls) decides its proportions. In Ilsan Residence these are 3.6m, larger homes in Seocho Residence are 4.5m. (Fig. 3) Modeul of the Kan creates the width of Chae and size of volume. The size of the space is decided by the size of each regular module, and at times, the addition of a corridor, or any other extended area from the Chae can add to and create the desired size of the space.

The way in which each Chae linked to each other was also an important topic for exploration. In those spaces where each Chae linked to one another, 'in-between' spaces would be made and those spaces were very interesting. They could become a hallway or small places of spare space, or connect to the living room and become a part of a wider central space.

The variety of ways in which each Chae can be placed allows a number of different layouts and so there was great flexibility in adapting to different types of site. When placed as multiple stories, it was possible to have open spaces that were two stories high and in this way, high volume spaces could also be created when necessary. The use of Chae and Kan to create spaces may be a universal architectural archetype, but I believed that depending on the shape of the land and the content of the programme, the strategy's responsiveness to the local environment allowed for specific expressions of local areas. Expressing locality while employing universal architectural vocabulary, and combining within the space the needs of the individual and the group is the greatest benefit of the mechanics of Chae and Kan.

Ilsan Residence:
Spatial Systems and the Narrative of Experience

The exploration of the Chae and Kan can best be seen in the Ilsan Residence. (Fig. 4) The Ilsan Residence was my first contract after starting my own practice. From the beginning of the project, my aim was to create a *hanok* for this era with modern architectural forms, and I hoped it would be the first project to realise an ambition that I had carried with me from my college days. On the one hand, my thoughts were of the traditional Korean *hanok* and on the other hand they were of the heritage of Modern architecture. And of course, my focus was on the way modern families lived.

First of all, I decided on three vault type volumes of 3.6m width, and created a plan for the spaces inside. The masses of three Chae, allowed for adaptability to any kind of terrain through utilising different placements. It is the living room and bedroom that is suited to the division of detailed components, and it was also well suited to creating different relationships between the indoors and the outdoors areas of the house. I wished to develop the sense of space that created by the different heights and surfaces of a Korean *hanok*, with a modern vocabulary within a single Chae. The living room with its exposed vaulted ceiling twice the height of a single story, the relatively lower and longer hallway, the single vault height of the family room, master bedroom and dining room, and the 75cm wide *tue-maru* in front of that, the balcony under the vault which looks onto Joengbalsan mountain, the living room connected to the cosy inner yard, all combine to create a varied space. The three separate Chae provided private spaces for the three generations to live with one another while at the same time also providing shared spaces in the living room and the court.

With only these three masses, it was not only possible to create an indoor space but also an expressive exterior for the outdoors, the road and the city. Although each mass was the same size, the direction to which it was turned, or a slightly angled placement, created a different view from any angle. The two-stories high entrance space, the protruding balcony and eaves, and the pilotis space created a surface for communicating with the city, and capable of expressing multiple urban relationships.

Chae and Kan were the building principles for the space, and the combination of Chae and Kan resulted in various different exterior spaces. Within the relationship between the interior and the courtyard, a variety of scales and spatial senses develop. The homes in the Ilsan Residence have an entrance yard, a courtyard and a service garden beside the kitchen, a yard beside the master bedroom; in total, four different courtyard, areas. The way each yard met with its indoor counterpart was different, and so although the house was small the experience is that of spaciousness. All the indoor spaces claim a piece of outdoor space in some way. The entrance with the entrance yard, the living and dining room with the porch and the courtyard, the master bedroom with the porch and the master bedroom yard, the kitchen and utility room with the service yard, the second story rooms also have balconies and terraces. All spaces are placed within the relationship of the public and the private.

I focused on the narrative structure of moving from the city to each road, each road to the entrance hall of the house, from the entrance hall to the living room, from the living room to the bedroom. (Fig. 5) The entrance gate, two stories high, connects to an entrance hallway, which is half that height, and the low ceiling of the entrance hall moves on once more to the living room where the ceiling once more rises to the height of two stories.

After moving through another one-story high hallway, and passing the front room, one connects to the bedroom. As the space develops, the size of the space and the amount of light accedes to and is adapted to the spatial scale of the city and the interior logic of the space. (Fig. 7) The nodal elements that decide the main space and the entrance space are drawn from both Eastern and Western architectural traditions. The development from the entrance of the land to the living room and the second-floor family room goes through several spatial phases, establishing a subjective experience of spatial narrative that is possible within an objective system of architecture like Chae and Kan. It is the containment of classic architectural vocabulary within a neutral system of architecture that is an evolution of Modern architecture. To me, architecture is the synthesis of construction systems and spatial narratives. Even the very simplest urban house contains the dimensions of both spatial systems and the narrative of experience. To become a universal archetype or model form, architecture must be a simple form within which an abundance of content can be placed.

The Architectural System for the Ilsan Residence, Development and Evolution

After putting together the proposal for the Ilsan Residence in 1994, I felt that it was clear that as a universal form, Chae and Kan could easily be adapted to different projects. Later, I developed similar ideas projects like the Dolce House and the Seocho Residence. (Fig. 8) By changing the size of the Chae I could set the spatial area that I wanted, and create cosy yard spaces between each Chae.

I felt that the architectural system of the Ilsan Residence could be applied to any building, not only that of residential projects. In 1995, the Ministry of Health and Welfare decided to make their construction units more uniform, and the Ilsan residential architectural system was nominated as the model standard during the design competition. (Fig. 9) The form utilised in the Ilsan Residence was one that can could utilised on any terrain and had a systematic form and structure. It took two years to complete local health care centres using the Ilsan Residence architectural system. The Health Care Center Chulwon, Health Care Center Gyeongju, Health Care Center Namhae are all prime examples of this.

This method of creating each Chae made by repeating the modules that form each Kan, and the building from the repetition of each Chae was developed further in the construction of the Ewoo School. The Ewoo School is composed of modular Kan measuring 8.4m × 1.8m. The placement of each separated school annex made it easy to adapt to the shape. (Fig. 11) The courtyards that then naturally appeared between each school annex became a feature of the school. The steel structure framework and drywall used in constructing the Ewoo School were factory-made prefabrications. It confirmed that it was possible to use modern production methods to construct the Chae and Kan form within a short time frame.

Single Mass – Box

If the Chae and Kan type employed in the Ilsan Residence was expanded for use when building Health Care Centers and Schools, the Single Mass type began life as Health Care Centers, and offices and then became adapted for residential projects. As universal systems, Chae and Kan and Single Mass types evolved through their adaptations to churches, medical facilities, schools and to other fields of architecture. If Chae and Kan began as an exploration of the traditional Korean *Hanok*, the Single Mass Box began as the archetypal form of the modern building.

As I worked on proposals using Chae and Kan for different buildings, including Health Care Centers and schools, a number of limitations became apparent. Although it was easy to create outdoor spaces through the development of Chae, during that process the size of the façade required would grow bigger, details would become more complicated and construction become more expensive. For public buildings like Health Care Centers and Hospitals, it is important to stay on budget and to use the Chae and Kan system made it difficult to meet financial goals. The Chae and Kan type is a single layered space and as such, is spatially shallow, making it difficult to create a sense of abundant space inside the building. Structurally it tends use single bays or single Kan, and so there are also limitations in reducing the number of supporting posts or beams. Furthermore, allocating and organising space to the many connecting passageways necessary between each Chae was also a challenge.

I had to find a solution other than Chae and Kan. I needed an ergonomic spatial form that could reduce both the total area required and the number of connecting passageways. After a lot of trial and error I arrived at the angular Single Mass form. The simple cube form made the surface area small, which made construction costs effective. Hallways and other public spaces could be created efficiently, so even with the same area size it was possible to produce a feeling greater spaciousness. Also, as the space became deeper it became possible to create a greater variety of spaces.

The Chae and Kan form used in traditional architecture is connected to the traditional structural system that made it difficult to create a long span. The large volumes presented by a steel frame structure and the reinforced concrete structures of modern architecture make it possible to place the combination of Chae and Kan into a single conformation. By creating a single large volume, it is possible to create a variety of spaces and levels within that. My exploration of the single mass box type with the discipline and process of architecture is, in this way, the inevitable result of the practical extremities with which I was met.

The First Single Mass – Box: Health Care Center Goseong

The first time I consciously applied the Single Mass – Box is the Goseong Health Care Center, constructed in 1998. The plan was nominated as the standard template in 1995, and it came after a few years of constructing a number of different public service buildings and struggling with keeping construction costs on target. As the Health Care Center Goseong budget was limited in the extreme, the idea of a simple cube shape was something I kept in mind from the very beginning of the process. My plan was to divide the simple cube shape into two floors, and create a wide opening at the centre of the building so that the whole building could be viewed in a single glance. The result was a significant reduction in construction costs and a more efficient use of space, while at the same time providing enough space to meet the needs of a public space. These forms also led to the single mass box type used in the Health Care Center Moongyeong and similar projects, continuing my exploration into all the attractions of a simple form.

The composition system of the Health Care Center Goseong contains principles that have evolved from the free-plan of modern architecture. As the columns stand independently of the exterior walls, the interior walls and the inner walls, both interior and exterior walls are excused from their load-bearing role and can be placed in any form. (Fig. 14) However, it was not my intention to apply the principles of free-planning to all spaces. In small spaces that needed to be further divided I combined the load bearing columns with the walls to maximize the amount of available space, while in larger, central areas I intended to have the load bearing columns separate from the walls to make the spatial logic apparent.

Ultimately, the principles of free-planning were not what was needed for the single mass form. This was because the catalysts for that beginning were to reduce construction costs by reducing the surface area of the building, making a deeper space to make the space more flexible and to create a more efficient public space.

Single Mass – Box Type as a House

If the use of the Single Mass – Box Type in public service buildings was devised by creating a modular system that focused on the spaces between load bearing columns, this kind of modular development was impossible for residential projects from the very beginning. Because houses are intrinsically much smaller, modular repetition is much more difficult, and it was further unfeasible to build rooms with special features, such as bedrooms and living rooms, or bathrooms and kitchens, as uniform modules. The solution that presented itself was to have the exterior walls of the house decide the structure of the house and to support this only when necessary within the house. It is a method that is quite different to that used by Mies Van de Roh's the Tugendhat House or in Le Corbusier's Villa Savoye where the columns are placed along a grid and the walls flexibly placed throughout. It is most similar to Adolf Loos' Villa Müller, but the structure of the space is slightly different. The Villa Muller creates spatial transformations through a series of rooms with different heights, while steps act as a mediator to connect each unique space. In the Gwacheon Residence, the S Residence, the Pangyo Pine Tree Residence, are all instances in which I experimented with the Single Mass type, I made a courtyard-like space in the middle of the house or opened up the focal living room to create a courtyard-like space. (Fig. 15) The contrast with Modern Architecture is that the courtyard structure of an urban house evolved from the Hanok and is placed within a cube volume.

Furthermore, unlike Western homes where the bedrooms are all grouped together, the master bedroom and the children's bedrooms are far apart, and the space between filled by a living room or a family room. This placement allows the family community to be the focal point of the house while still providing the traditional physical and psychological space between family members as seen in the *hanok*.

Although design began as a simple cube, the shape became more complicated with the addition of a courtyard and balcony. Ultimately, a house is a place in which multiple relationships are formed, between person to person and city to nature, and so the system cannot help but become more intricate.

Gwacheon Residence was requested as a solution to satisfy the requirements of building on a limited area of land, as economically as possible. It is composed of a cube that is two-stories high and has a sloping roof. (Fig. 16) The whole framework is composed of exterior walls and is made of thin walls that are load bearing and a load bearing column in the walls behind the first floor stairs. As the supporting column is of minimal thickness and is tucked in between the walls the column cannot be seen from within.

The cube-shaped space is split into two and the living room area is open to the second floor. In the Gwacheon Residence, the living room functions as an inner courtyard and is the focal point of many activities. Arranged around the perimeter of the living room are the dining room and kitchen, master bedroom and study, and the children's rooms, just as if the living room were the courtyard of a urban traditional *hanok*. (Fig. 18) When placing a house within a cube shape, the grid of columns disappears into the walls and the order of each Kan becomes almost invisible. It is simply that the living room or the courtyard becomes the central area and the whole space is segmented into 3 × 3 or 3 × 2.

The S residence contains a variety of different spaces within the precise angles of the concise shape of a cube, where three generations live together. (Fig. 19) While each generation has their own unique area, their methods of communication, and (beyond the single mass shape) the methods of communication between the indoors and outdoors, the different levels and volumes within the single mass box that create a spatial sense of leisure, the four different courtyards that have been brought inside the Single Mass type, there is a large repertoire of different spaces within the S Residence. The very simplicity of the lines of the cube enables the spatial flexibility and freedom there within.

The clients wanted a structure that could easily adapt to changes in family members, and this was produced through a spatial system and a structural system. The angles of the cube shape of the exterior and the smaller cube in the interior decide the structure. Between the two cubes is a space that can flexibly change its functionality depending on future changes. In the inner small cube are stairs, an elevator and a courtyard. Although it is a small area it combines multiple functions. By climbing up and down the stairs and passing through the central space, it is possible to have a variety of spatial experiences and to create an abundant architectural sense of place. The inner cube functions like the courtyard of a square urban house, and the other spaces of the house are developed around its perimeter. Around the square of this central space, the demarcated spaces of the living room and dining room, the entrance hallway, the kitchen and service room, and the master bedroom are arranged. Each is separated as an independent Chae, but they are gathered and placed neatly together inside a single cube.

Starting from the simplest type of a box, the **Pangyo Pine Tree Residnece** is an amalgamation of the principles created from the clients' requests and the conditions of the site and its neighbours. (Fig. 21) Although it is a simple box shape, the diversity of daily life meant that the shape of the box naturally evolved. The living room, kitchen, bedroom, attic and so on have different ceiling heights which produce a variety of levels. Separated levels are projected onto the façade and create various expressions of window treatments which subtly change the appearance of the cube. The courtyard, with its unique personality, is inserted inside the box subtly transforming the appearance of the box still further and continuing its interesting evolution. The second-floor balcony from which one can look out at the view, and the large void that faces the linear stairs that leads to the second floor, the courtyard intended as a private space for the clients, the attic and the connecting rooftop garden are all examples of different spaces that exist within this cube shaped box.

As the spaces within the box are used for multiple functions, the house becomes a place where different worlds co-exist, where it becomes apparent that different existences are together. Areas are combined and connected in ways that eliminate wasted space. The intention to create a form where different things can exist as they are, without changing but existing alongside each other, as efficiently as possible.

Type – Universal Systems and Specific Architecture

The architectural theory that exists only as an alibi for the architect cannot survive for long. It is only when it is born as a reflection of the times and of the place and the universality of those things that it can be understood, criticized and finally, evolve. Architecture is not simply the works of any single architect or the property of any single person, but exists as the inheritance of the community. Therefore, in order to look for methods of communication for society one must search for the enabling fundamental principles within architecture.

Exploration of shape (type) is both the theoretical methodology of the attempt to find the fundamentals of a universal theory and its practical application. The types presented by Chae and Kan, and the Single Mass – Box contains within the name the principles for producing architecture. Within that, the principles for spatial relationships, structural systems and the management of shapes are also contained. Here, type can be defined as the grammar that creates space. The discipline and process of architecture must encompass the universality and flexibility of many years of architecture, and the systems that it needs. It must also contain the possibility for architectural projects to be re-interpreted and continue to evolve.

I believed that it was necessary to look at Western Modern architecture and contemporary architecture, but rather than simply passively receiving that architectural vocabulary, we should understand it and question it in the context of the way we live, and develop it. New programmes and new sites always demand special solutions, but on closer examination one can see overt universal values and conditions for sharing. Each cooked dish is unique but the plate ware that serves those dishes or the order of courses can be universal. The exploration of form may look like an attempt to find a single answer to architecture, but I would argue that it is actually the opposite. Through the exploration of a grammar that produces architecture we can build bridges to discover the characteristics and identity of our generation and our locale. With those bridges we can create architecture that contains our unique preferences and vision.

유형을
넘어서

Beyond Types

탐구의 다양한 주제들

'채와 간', '단일매스 박스' 유형이 공간의 체계에 관한 것이었다면, 우리 시대가 만들어내는 다양한 조건, 특히 사회-경제-도시계획-부동산 개발로 이어지는 여러 상황들은 또 다른 차원의 주제를 생산한다. 새로운 주제들은 다양한 거주의 요구와 함께 등장했다. 2000년 이후 단독주택 프로젝트 의뢰가 급격히 많아졌고, 그 조건이 다양해졌다. 그것은 단독주택이라는 형식이 그 조건에 따라 다양하게 분화되고 있음을 의미했다. 제기된 주제를 몇 가지로 분류하면, 판교 등 신도시에 조성된 '신도시주택', 시골의 넓은 대지에 지어지는 '전원주택', 소위 '타운하우스'로 불리는 '마을-주택'으로 크게 나눌 수 있다.

신도시의 주택들: 공공성과 개인의 취향

신도시 주택지의 풍경과 단독주택의 공공성

판교 등 2기 신도시 단독주택지의 도시설계 지침은 1990년대 만들어진 1기 신도시의 경우와 큰 차이가 없다. 비슷비슷한 크기의 주택 필지, 직교좌표를 따라 만들어진 길, 평평하게 조성한 지형, 그 어떤 특징도 없는 국적불명의 마을 계획이다. 단독주택 블록의 몰개성은 개별 주택의 과도한 개성으로 귀결된다. 결국 마지막에 우리가 신도시 마을에서 만나게 되는 것은 개인의 욕망이 최대로 실현된 난개발된 풍경, 주택의 동물원이다.

Fig. 24 판교 신도시 위성지도
Pangyo New Town map

Fig. 25 고양이집(2014) / 판교의 주택지 역시 다르지 않은 조건이라, 주택지의 획일적인 바탕 위에 제각기 다른 주택들로 채워질 수밖에 없는 상황이었다.

House Cat (2014) / As the houses in Pangyo were also subject to the same conditions, the only option was to try and fill each uniform lot of the residential areas with different types (shapes) of housing.

판교의 주택지 역시 다르지 않은 조건이라, 주택지의 획일적인 바탕 위에 제각기 다른 주택들로 채워질 수밖에 없는 상황이었다. 판교의 도시설계 지침은 건축선, 녹지축 등 일산 신도시보다는 강화된 통제수단을 갖고 있지만, 반드시 있어야 할 높이에 대한 규제가 없고, 조경이나 건축재료에 대한 규정은 너무 약했다. 거기에다 도시주택에서 프라이버시를 지키기 위해 필요한 건축적 장치인 담장과 차폐 조경을 금지했다. 이는 미국 교외의 주택지처럼 넓은 대지 위에 집이 띄엄띄엄 배치되는 경우에나 작동하는 지침이다.

2008년, 판교 11블록 건축주 인터넷 카페 회원들이 내게 주택설계를 의뢰했다. 그들은 개별 주택도 중요하지만 전체 블록이 조화로운 풍경을 이루기를 원했다. 난개발된 신도시 주택가의 풍경이 여기서 되풀이되면 안 된다는 것이 모두의 생각이었다. 11블록의 코디네이터-건축가로서 조화로운 경관과 프라이버시 보호를 위해 지켜야 할 몇 가지 가이드라인을 제시했다. 가이드라인은 법적 효력이 있는 것은 아니지만 기존 도시설계 지침에서 다루지 않은 핵심적인 사항들을 포함하고 있다. 그 내용은 주택의 최고 높이를 제한하는 것, 이웃 간에 프라이버시를 지키기 위한 장치를 도입하는 것, 조경의 일관성을 유지하는 것, 외벽재료의

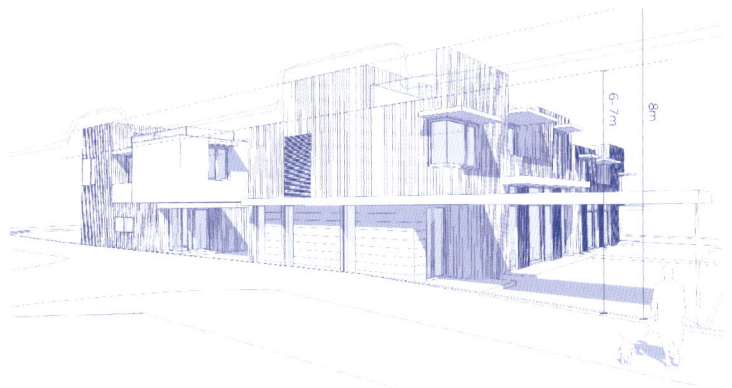

Fig. 26 판교 11블록 가이드라인 / 가이드라인은 법적 효력이 있는 것은 아니지만 기존 도시설계 지침에서 다루지 않은 핵심적인 사항들을 포함하고 있다.
These guidelines had no legal power but they dealt with issues that were not covered by the existing city regulations.

종류를 제한하는 것 등으로 요약할 수 있다. 하지만 금융위기가 도래하면서 대다수 참여 주민들이 착공을 유보했고, 협의체에 속하지 않은 대지 소유자들은 가이드라인을 따르지 않았다. 결국 모든 노력은 실패로 귀결되었다.

그럼에도 그 과정에서 신도시 주거블록 전체 풍경을 조화롭게 만들자는 건축주들의 목표를 이해하게 되었고, 이후 10년 동안 20개 프로젝트로 이어진 신도시 주택 연작에서 핵심적인 관점을 지키려 노력했다. 그것은 주택이 갖추어야 하는 '공공성'이다.

집은 도시의 일부를 구성하는 일원으로서 적절한 높이와 크기를 가짐으로써 도로와 이웃의 집이 적절한 공간 비례로 만들어질 수 있게 해주어야 한다. 2층의 적정 높이인 6~7m를 지키되, 다락이나 옥탑 등 꼭 필요한 경우에 한해 부분적인 돌출을 허용한다. 이웃의 프라이버시를 지키기 위한 건축적 배려도 중요하다. 이웃의 마당 쪽으로는 창을 최소한으로 설치하고, 창에는 그릴 방범창 등으로 시선을 차단하는 장치를 반드시 설치한다. 공통된 수종의 식재를 통해 동네의 성격을 만들되 개인의 취미가 반영된 특별한 수종의 식재를 허용한다. 노출콘크리트, 목재, 자연석 등 재료 본래의 맛이 살아있는 재료를 구사하도록 했다.

마을 조성에서 '공공성'이란 목표는 근사한 문장이나, 뜨거운 구호로 성취될 수 없다. 마을에 대한 성찰과 비전을 바탕으로 계획한 마스터플랜, 마을의 풍경과 도시주택에 대한 깊은 이해를 담은 가이드라인, 그리고 이에 대한 건축가와 시민의 존중으로 완성된다.

개별 주택 단위의 요구들

이 시대 주택에서 요구되는 것은 주거 공간을 통한 자아의 실현이다. 이제, 집은 단순히 방과 거실, 부엌의 조합이 아니다. 가족구성원들 각자가 갖고 있는 서로 다른 관심사, 서로 다른 취미를 집이 담아낼 수 있어야 한다. 그것은 서로 다른 특성을 갖는 공간들이 하나의 집의 윤곽 속에서 배치된다는 것을 의미한다. 따라서 건축가는 집의 윤곽을 만들어내면서도 '다른 취향'을 공간적인 장치로 담아내야

한다. 그 결과 '차이'와 '다름'은 중요한 목표가 되었다. 서로 다른 것들이 모여 하나의 집을 이루어야 하며, 다른 것들을 하나로 엮기 위해 여러 해법이 시도되었다. 재료의 종류를 제안하는 것, 다름을 형태적, 공간적으로 드러내는 것, 하나의 윤곽 아래 서로 다른 공간을 담는 등, '차이'를 건축적으로 조율하는 방식에 따라 주택의 특징이 자연스럽게 만들어진다.

    이 시대 건축주들의 중요한 특징은 왕성한 정보 수집과 교환이다. 인터넷 카페를 통해 설계와 시공에 관한 여러 정보를 얻는다. 한편, 인터넷 서핑을 통해 주택의 이미지를 파편적으로 수집한다. 그들이 수집한 정보나, 인터넷 서핑으로 구한 이미지의 리스트가 건축가에게 참고가 될 때도 있지만 외려 설계에 방해가 되는 경우도 있다. 기후와 지형, 공사비, 대지의 크기 등, 그 출발점이 다른 주택의 특징과 이미지가 현실에 구현되기란 어렵기 때문이다.

    신도시의 주택을 설계하는 건축가는 두 개의 극단을 만난다. 지극히 무성격한 대지와 특별한 집에 살고자 하는 건축주의 열망, 그 극단의 거리를 조율하는 것이 건축가의 역할이다.

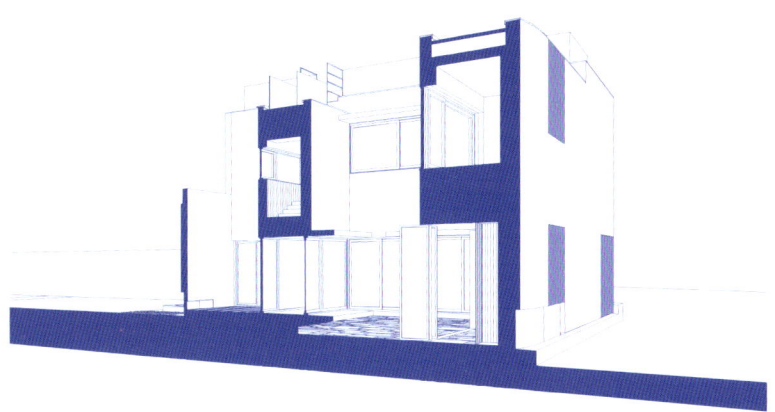

Fig. 27 손톱집 단면 다이어그램 / 손톱집의 입체는 밖으로 닫히고 안으로 열려 있어, 주변의 고층 아파트로부터 내부의 프라이버시를 지켜준다.
House Nail section diagram / By creating a shape that is closed off from the outside but is open on the inside remains private from the surrounding high-rise apartments.

손톱집은 직육면체의 윤곽이 대지의 형상에 맞게 변형된 프로젝트이다. 손톱처럼 생긴 평면에 세 개의 마당이 삽입되면서 공간이 구성된다. 현관 앞마당은 도시와 이어지는 접촉면으로 주차장과 이어져 진입 공간을 형성한다. 거실-식당 앞 중정은 집의 중심이 되어 부엌과 식당, 침실 등 모든 공간과 만난다. 한편 거실과 계단 사이의 작은 마당은 거실의 풍경을 위한 공간이다. 손톱집은 마당이 주인공인 집이며, 마당을 중심으로 공간이 전개되고 마당을 향해 풍경이 만들어진다.

    손톱집은 건물의 외곽을 만드는 외벽면과 내부의 중정을 만드는 외벽면의 차이가 두드러지는 주택이다. 건물의 외곽은 검은색 벽돌로 마감되고, 창문이 차지하는 면적이 작아서 폐쇄적인 느낌을 준다. 반면에 중정의 외벽은 목재로 만들고, 넓은 창문이 많아 개방적인 공간감을 준다. 손톱집의 입체는 밖으로 닫히고 안으로 열려 있어, 주변의 고층 아파트로부터 내부의 프라이버시를 지켜준다.

Fig. 28 건물의 외곽은 검은색 벽돌로 마감되고, 창문이 차지하는 면적이 작아서 폐쇄적인 느낌을 준다.
The exterior walls of the building are finished with black brickwork, and the small windows have a closed off and protected privacy.

Fig. 29 중정의 외벽은 목재로 만들고, 넓은 창문이 많아 개방적인 공간감을 준다.
In contrast, the walls of the inner courtyard are made of wood, and the many wide windows give a sense of freedom and shared space.

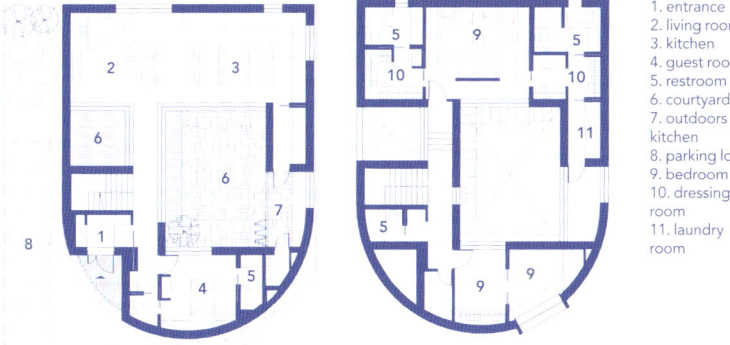

1. entrance
2. living room
3. kitchen
4. guest room
5. restroom
6. courtyard
7. outdoors kitchen
8. parking lot
9. bedroom
10. dressing room
11. laundry room

Fig. 30 손톱집 평면도 / 손톱집은 마당이 주인공인 집이며, 마당을 중심으로 공간이 전개되고 마당을 향해 풍경이 만들어진다.
House Nail plan / The yards are the main attraction of the House Nail. The spaces are developed around these yards and while the landscape is created by facing these yards.

Fig. 31 판교 ㄷ자 집은 중정을 중심으로 공간이 전개되면서 모든 공간은 마당과 밀접한 관련을 맺는다.
The space unfolds, centred on the courtyard, keeping an intimate relationship with the garden.

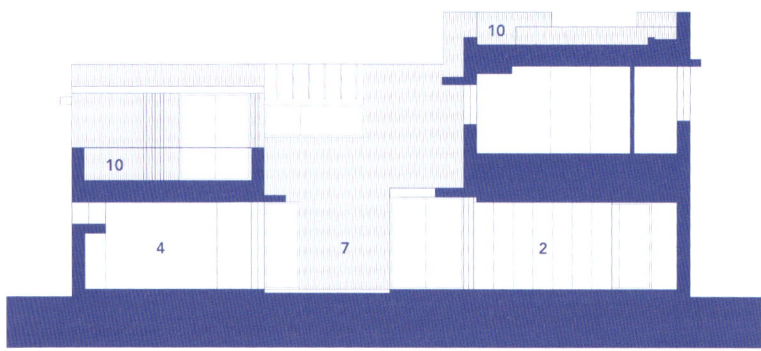

Fig. 32 판교 ㄷ자 집 단면도 / 중정에 면한 처마와 발코니, 창과 벽, 테라스와 옥상정원 등 다양한 건축적 장치를 통해 기후에 반응한다.
Pangyo ㄷ-shaped Residence section / Architectural apparatus, such as eaves and balconies, windows and walls, terraces and roof gardens, are installed to respond to the climate, providing unique features and organising narratives of experience.

주택의 특별함이 마당에서 비롯된다면, 도시주택의 주인공은 중정이다. 도시주택이 중정을 가질 수밖에 없는 것은 좁은 대지에 가장 효율적으로 내밀한 외부 공간을 만들 수 있기 때문이다. 이 **판교 ㄷ자 집**은 중정을 중심으로 공간이 전개되면서 모든 공간은 마당과 밀접한 관련을 맺는다. 중정의 한쪽 벽이 가변적이어서 중정이 열리기도 하고 닫히기도 하며 도시와의 관계를 필요에 따라 조절한다. 또한 중정에 면한 처마와 발코니, 창과 벽, 테라스와 옥상정원 등 다양한 건축적 장치를 통해 기후에 반응한다. 각 공간은 고유한 특성을 부여하며, 경험의 서사를 조직한다. 하나의 윤곽 안에 개개인의 특별한 방이 배치되어 있다. 음악실, 식당, 거실, 자녀방 등 각각의 방은 자기 자신만의 외부 공간을 향유한다.

Fig. 33 판교 ㄷ자 집 평면도 / 음악실, 식당, 거실, 자녀방 등 각각의 방은 자기 자신만의 외부 공간을 향유한다.
Pangyo ㄷ-shaped Residence plan / Although configured along a single contour, each room enjoys its own exterior space.

판교에서는 드물게 경사지에 위치한 **자오당**은 땅 아래의 질서와 땅 위의 질서가 만나는 곳이다. 마당을 주변으로 내부 공간의 체계를 짜고, 도시적 관계를 담은 주택의 경계면을 구축해야 했다. 바깥어른의 서재는 도시와 중정 사이에 개방적인 포즈로 위치한다. 한옥의 사랑채에 해당하는 서재에서는 안과 밖을 동시에 조망하고, 내부와 외부의 공간을 아우른다. 거실은 마당 깊은 곳에 있는 아늑한 공간이다. 크기와 성격이 다른 두 개의 중정이 거실과 식당에 고유함을 부여한다. 2층에 위치한 세 개의 방은 서로 다른 단면의 형상을 갖는다. 같은 지붕 아래 있지만 각각의 방은 특별하고 또한 유일한 공간의 형태와 조직을 갖는다. 그것은 가족과 개인의 관계와도 같다. '집'은 하나의 집이자 동시에 여러 개의 집이다.

Fig. 34 판교에서 드물게 경사지에 위치한 자오당은 땅 아래의 질서와 땅 위의 질서가 만나야 하는 땅이다. Unusually for Pangyo, the land of House Ja-O-Dang on which this house is built is on a slope, and so it must meet the requirements of the land both above and below and it.

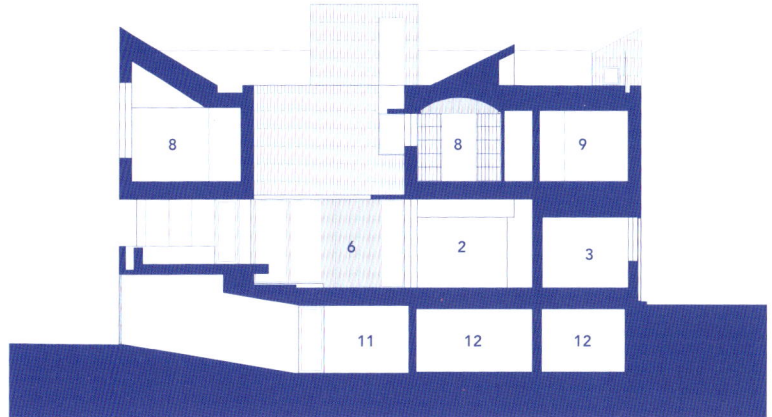

Fig. 35 자오당 단면도 / 한옥의 사랑채에 해당하는 서재에서는 안과 밖을 동시에 조망하고, 내부와 외부의 공간을 아우른다.

House Ja-O-Dang section / Reflecting the role of the *Sarangchae*, the study has views of both inside and outside the house, and forms a meeting place for the indoors and outdoors.

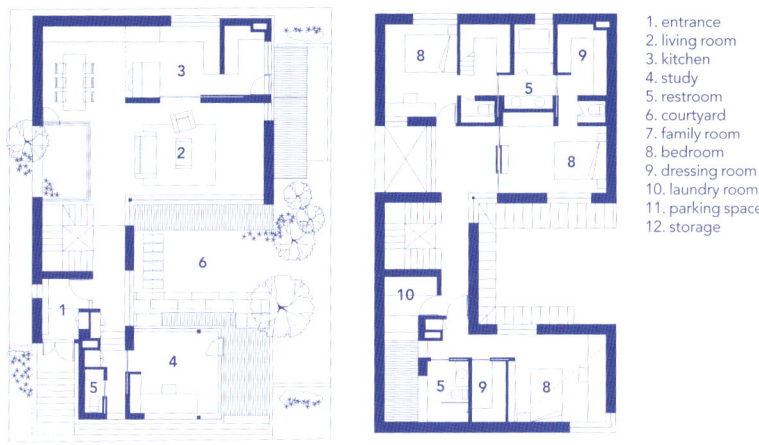

1. entrance
2. living room
3. kitchen
4. study
5. restroom
6. courtyard
7. family room
8. bedroom
9. dressing room
10. laundry room
11. parking space
12. storage

Fig. 36 자오당 평면도 / 같은 지붕 아래 있지만 각각의 방은 특별하고 또한 유일한 공간의 형태와 조직을 갖는다.

House Ja-O-Dang plan / Although they are all under the same roof, each room has its own unique shape and organisation.

전원주택의 실험:
마당과 방, 그 소우주의 풍경

2000년 이후 소득이 향상되면서 전원에 주택을 갖고자 하는 사람들이 크게 늘었다. 별장으로 우선 사용하다가 은퇴 후에 자신의 거처로 쓰려는 수요가 많았다. 도시주택과 달리 전원주택은 대지가 넓고 주변에 아름다운 경관을 갖고 있다. 외부 공간이 적절히 분화되고, 마당을 내부 공간과 관계를 맺으면서 유기적으로 구성하는 것이 요구된다.

 한옥은 주택의 배치, 마당과 내부 공간의 전개를 탐구할 때 좋은 선례가 되었다. 건물과 담장을 적절히 구사해서 넓은 대지를 여러 영역으로 나누고, 행랑채, 사랑채, 안채, 정자 등으로 주택의 내외부 공간에 분명한 성격을 부여하는 방식에 주목했다. 특별히 소쇄원의 정원과 건축이 교훈이 되었다. 직교좌표로 짜인 담장과 축대, 건물의 배치를 통해서 공간의 영역을 짜임새 있게 구축하는 동시에, 주변의 풍광을 대지 안으로 끌어들이는 구성을 배울 수 있었다. 한편 중국의 정원건축도 좋은 참고가 되었다. 좁고 긴 진입 공간을 지나 넓게 열린 정원을 만나는 서사 구조가 인상적이다. 특히 소주 유원의 공간 구성이 자주 떠오른다.

 건물로부터 마당을 향해 뻗어가는 벽과 담장을 통해 외부 공간을 제어하는 방식은 미스 반 데어 로에의 브릭하우스 계획안, 프랭크 로이드 라이트의 유소니언 하우스 연작에서도 쓰인 방식이다. 근대건축의 사례에 주택 정원의 분화와 전개가 보이지만, 전체적으로는 정원이 배경이 되고 건물이 주인공이 된다. 이에 반해 한옥은 외부 공간이 쓰임새에 따라 분명히 정의된다. 또한 건물은 하나의 오브제로 읽히기보다는 분리된 마당에서 만난 여러 모습의 콜라주(colllage)로 인식된다. 건물은 마당에 따라 다르게 존재하기 때문이다.

 주택 프로젝트의 규모가 비교적 작고, 내부 공간이 하나로 이어져야 했기에 전통한옥과 같은 공간의 전개에 한계가 있었다. 그렇지만 담장과 내부 공간의 분

Fig. 37 소쇄원 배치도
Soswaewon Garden site plan

절을 통해 건축과 마당의 콜라주를 만들 수 있었다. 설계의 첫 단계는 주택의 경계를 정의하는 일이다. 담장이 경계를 만드는 가장 중요한 수단이 되었다. 담장은 경계를 이루는 장치를 넘어 다양한 기능을 수행했다. 마당을 정의하고 건물과 만나 새로운 장면을 만들었다. 경사를 정리하기 위해 만든 옹벽이나 석축은 경계가 되고, 숲과 만나는 지점에는 숲 그 자체가 경계가 되었다. 담장과 축대와 수목이 적절히 경계를 구성할 때 주택의 영역은 주변과 조화를 이루었다. 그 모든 노력 끝

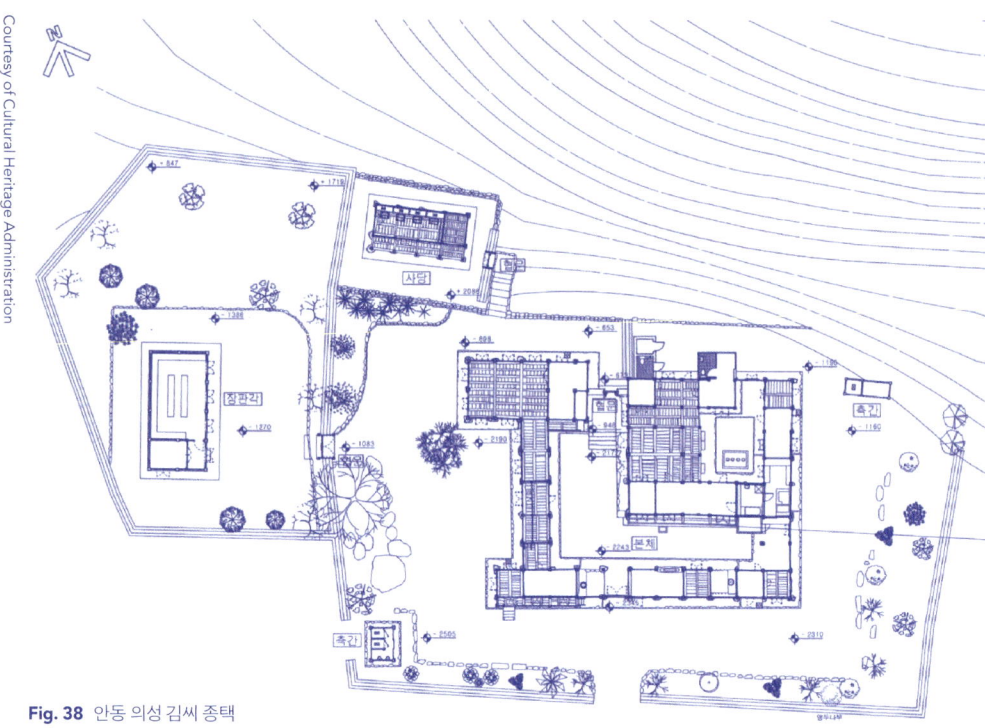

Fig. 38 안동 의성 김씨 종택
Head House of the Uiseong Kim Clan, Andong

에 만들어진 집은, 그 자체로 주변의 풍경과 잘 어울리는 또 하나의 새로운 풍경이 되었다.

대지가 넓은 만큼 진입의 과정을 충실히 만들어야 한다. 주차장부터 입구 마당, 현관에 이르기까지, 여러 단계의 공간을 지나며 환영 의식을 치른다. 무엇보다 중요한 것은 내부 공간과 마당의 결합이다. 주택의 각 공간을 나누어 다양한 종류의 마당을 만들었다. 넓은 대지에 자리 하는 전원주택은 작은 소우주를 지향하고, 그 우주는 건물과 담장, 조경 등 여러 켜의 공간적 장치에 의해 보호된다. 겹겹이 만들어지는 마당과 방의 결합을 통해, 거주를 위한 소우주가 완성된다.

양평주택은 비교적 긴 형상의 대지에 집을 펼치면서 여러 마당을 만들어낼 수 있었다. 거실, 식당, 부엌, 안방, 손님방 등 각 공간을 최대한 분산 배치하고 그 사이를 마당으로 채웠다. 어쩌면 마당을 먼저 설정하고 그 사이를 내부 공간으로 채웠다고도 할 수 있다.

마당이 하나의 레벨로 정리된 것처럼 보이지만 미묘하게 그 높이가 서로 다르다. 높이뿐 아니라 마당의 재료, 마당의 크기도 서로 다르다. 마당의 차이를 만든 것은 그 마당이 만나는 내부 공간의 성격이다. 식당 마당의 바닥과 벽은 식당과 같은 마천석과 콘크리트로 마감되었다. 안방 마당이 목재로 마감된 것도 같은 이치이다. 마당과 방의 세트는 각자 고유한 기능과 특징을 담고 있다. 거실이 내부 공간의 중심이듯이 거실 마당은 마당의 중심이다. 침실이 가장 은밀해야 하듯 침실 마당도 그렇다. 마당은 집의 배경이면서도 주제이다. 그것은 여러 성질의 물질로 이루어진 이 집의 담이 집을 구축하는 수단이자 배경이면서, 또 다른 독립된 주제인 것과 마찬가지이다. 건축물과 공간, 마당과 담, 방과 마루, 서로에게 기대면서 존재하는 것이다.

Fig. 39 양평주택은 비교적 긴 형상의 대지에 집을 펼치면서 여러 마당을 만들어낼 수 있었다.
The Yangpyeong Residence is situated on a comparatively long piece of land and so it was possible to create multiple gardens.

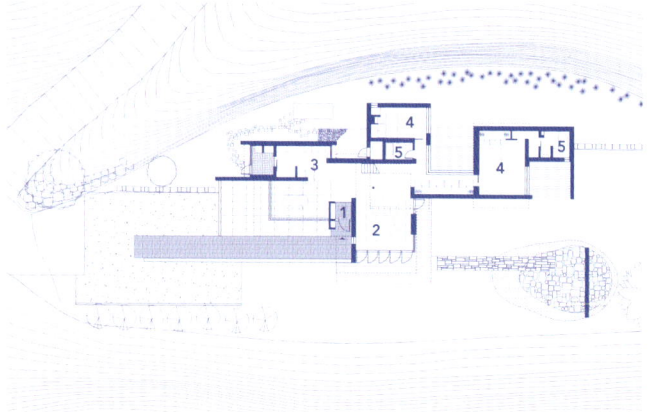

Fig. 40 양평주택 평면도 / 거실, 식당, 부엌, 안방, 손님방 등 각 공간을 최대한 분산 배치하고 그 사이를 마당으로 채웠다.
Yangpyeong Residence plan / The living room, dining room, kitchen, master bedroom and guest bedrooms were given as much separation as possible and a garden was placed between each of these rooms.

Fig. 41 마당과 방의 세트는 각자 고유한 기능과 특징을 담고 있다.
The matching gardens and rooms each have unique features and functions.

봉가리주택은 '집은 집들이다'라는 관점이 가장 분명하게 표현된 집이다. 여럿이 모여 하나의 가족을 이루듯이 '집들'이 모여 '집'이 된다는 생각이 담겨 있다. 주택의 각각의 공간을 독립된 또 하나의 '집'으로 만들었다. 집과 집 사이로 길이 생기고 마당이 생긴다. 집은 작은 마을이고 도시이다. 골목이 있고, 광장이 있다. 집 안에 담긴 '집들'은 각자 자신만의 고유한 공간과 마당을 갖는다. 각각의 집은 프로그램, 규모, 오리엔테이션이 다르지만 공통된 재료와 디테일이 사용되면서 가족으로서 유사성을 갖는다. '집들'이 모여 '집'이 된 이 주택은 집을 만드는 새로운 관점을 제안한다.

1. entrance
2. living room
3. kitchen
4. bedroom
5. dressing room
6. restroom
7. courtyard
8. parking lot
9. library
10. family room

Fig. 42 봉가리주택 평면도 / 집들이 모여 집이 된 이 주택은 집을 만드는 새로운 관점을 제안한다.
Bongga-ri Residence plan / The Bongga-ri Residence thus offers a novel perspective of building an archetypal house, conceived in its gathering of separate houses.

Fig. 43 가평주택은 남쪽으로 열린 전망을 누릴 수 있도록 동서 방향으로 길게 공간을 전개했다.
The Gapyeong Residence is developed in an East-West direction so that the southern aspect of the house can be viewed.

가평주택은 급경사지라 불리한 점도 있지만 지하 주차장을 도로 레벨과 같게 만들 수 있고 마당과 실내에서 멀리까지 트인 전망을 누릴 수 있는 장점이 있다. 남쪽으로 열린 전망을 누릴 수 있도록 동서 방향으로 길게 공간을 전개했다. 지하 주차장에는 선큰 마당을 두어 환한 빛이 계단을 따라 들어오도록 했고, 주차장에서 거실에 이르는 과정을 흥미롭게 조직하는 것이 중요한 과제였다. 지하 주차장에서 햇살이 쏟아지는 선큰 마당을 지나 멋진 계단을 오르면, 마침내 너른 마당을 지나 동서로 길게 펼쳐진 집을 만난다. 거실과 주인침실은 독립된 영역으로 구별되어 위치하고, 각 공간에서 고유한 마당의 전경을 누릴 수 있도록 했다. 경사지붕 아래 홍송으로 마감된 공간은 멀리 풍경을 감상하며 사색하고 휴식하는 장소이다. 히노끼 욕조가 마련된 옥상데크에서는 오직 푸른 하늘만 보인다. 고요한 공간에서 마주하는 하늘과 구름, 바람과 비가 오랜 세월 힘써 일한 건축주에게 선물처럼 다가오기를 바랐다.

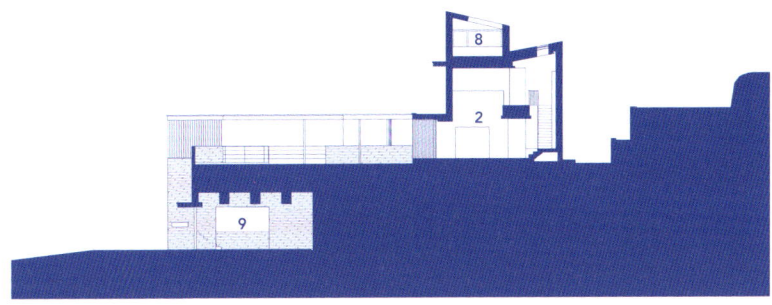

1. entrance
2. living room
3. kitchen
4. study
5. bedroom
6. dressing room
7. restroom
8. terrace
9. parking lot

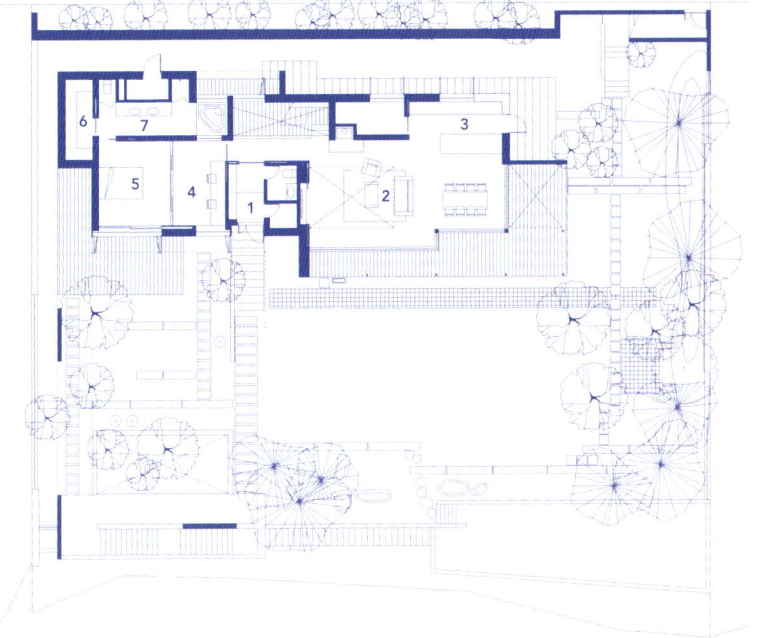

Fig. 44 (위) 가평주택 단면도, (아래) 가평주택 평면도 / 거실과 주인침실은 독립된 영역으로 구별되어 위치하고, 각 공간에서 고유한 마당의 전경을 누릴 수 있도록 했다.
(top) Gapyeong Residence section, (bottom) Gapyeong Residence plan / The living room and master bedroom are placed in an independent space, and each room enjoys a private garden.

여주주택은 건축가인 나 자신의 작업실이다. '집은 집들이다'라는 집에 대한 나의 입장이 담겨 있다. 집이 존재를 담는 장치라면, 존재가 지닌 복수성에 의해 결국 집은 '집들'이 될 수밖에 없다. 여주주택은 하나의 집이지만 40평이 안되는 아담한 면적 안에 서로 다른 성격과 공간감을 가지며, 또 서로 다른 단면을 지닌 작은 공간들이 담겨 있다. 1.9m 천장고를 가진 욕실 겸 마루, 두 층 높이로 열린 거실, 2.3m 큐브의 작은 방, 경사지붕 아래의 서재, 좁고 낮은 다락방 등, 11개의 서로 다른 단면 공간이 필요에 의해 생겨났다. 작은 공간들은 작은 스케일로 만들어진

계단, 가구, 디테일에 의해 풍부한 공간감을 획득한다.

    이 집은 다양한 모습으로 변화한다. 분절된 각 공간들은 슬라이딩도어, 한지 창문 등의 건축적 장치를 통해 열리고 닫히며 계속 변화하는 공간을 만들어낸다. 이 집은 계절과 시간에 따라 원하는 모습으로 변화한다. 집 안에 담긴 11개의 서로 다른 공간들은 건축적 장치와 외부 공간을 매개로 모든 시간마다 고유한 모습을 갖는다.

Fig. 45 작은 공간들은 작은 스케일로 만들어진 계단, 가구, 디테일에 의해 풍부한 공간감을 획득한다.
Small spaces use stairs built to a smaller scale, furniture and details create a greater sense of space.

1. entrance
2. living room
3. kitchen
4. bedroom
5. bathroom
6. restroom
7. terrace
8. backyard
9. vegetable garden
10. parking lot

Fig. 46 여주주택 평면도 / 이 집은 계절과 시간에 따라 원하는 모습으로 변화한다. 집 안에 담긴 11개의 서로 다른 공간들은 건축적 장치와 외부 공간을 매개로 모든 시간마다 고유한 모습을 갖는다.
Yeoju Residence plan / This house changes according to the season and the time of day. The eleven different spaces within the house are mediated by architectural devices and the outdoor spaces and every hour takes on a new aspect.

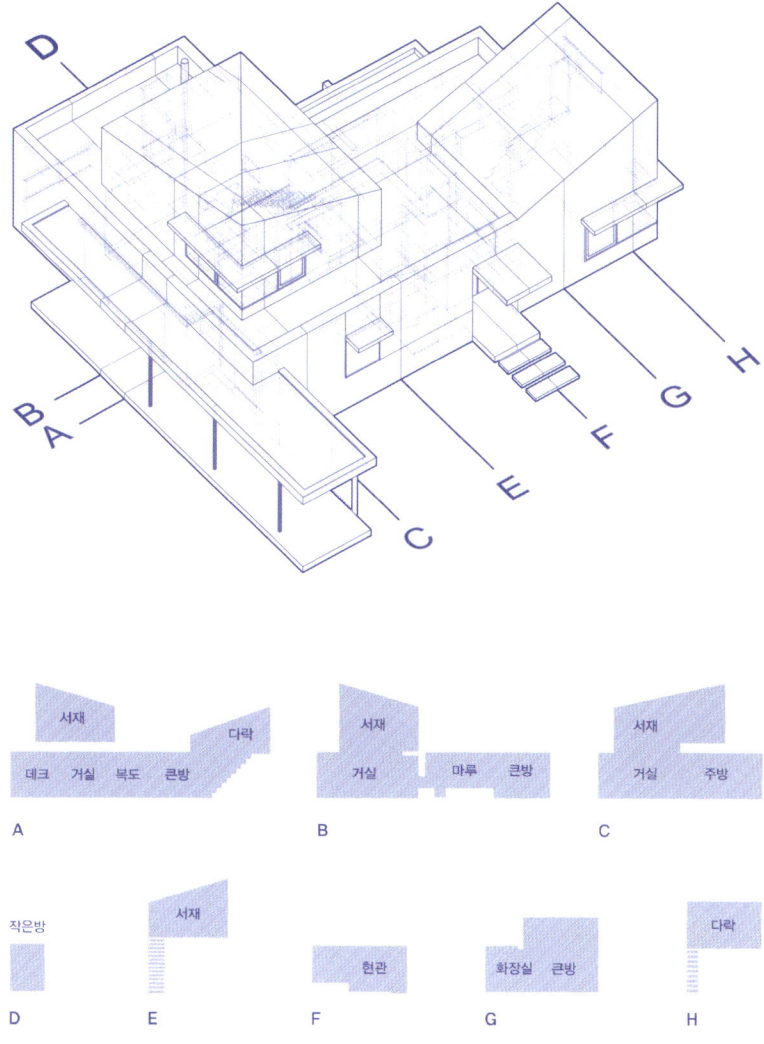

Fig. 47 여주주택 단면 다이어그램 / 여주주택은 하나의 집이지만, 서로 다른 성격과 공간감을 가지며, 서로 다른 단면을 지닌 작은 공간들이 담겨 있다.

Yeoju Residence section diagram / The Yeoju Residence is a single house but within the cozy area, contains many spaces that have different surfaces and characters.

타운하우스와 마을-주거:
집합의 문법 찾기

탐구의 주제로 새롭게 등장한 것은 여러 채의 주택이 공동으로 개발되는 주거다. 2000년대 중반을 지나면서 개발업자들이 단독주택 보급에 나서기 시작했다. 이들은 대지 구입-설계-시공을 하나의 패키지로 만들어 공급했다. 주택을 구입하는 사람은 번거로운 과정을 덜고, 개발업자는 싼값에 대량으로 구입한 대지에 부가가치를 더해 이익을 창출했다. 이런 주택을 '타운하우스'라는 이름을 붙여 공급했는데, 영국과 미국의 타운하우스가 우리나라의 연립주택, 즉 저층 집합주택에 가깝다. 공동개발함으로써 시공비를 절약하고, 커뮤니티의 공동시설, 관리실 등이 함께 조성되어 안전과 관리에 유리한 점이 많아 한동안 인기를 끌었다.

타운하우스의 설계는 공동개발이 전제이므로 디테일과 건축재료의 표준화가 요구되었다. 그보다 더 중요한 것은 평면의 개발이었다. 제시된 단독주택의 평면이 불특정 다수의 건축주에게 설득력 있게 받아들여져야 하기 때문이다. 단독주택의 평면을 개발하는 과정에서 도시주택의 유형을 고민하고 실제로 적용해 보았던 지난 경험이 크게 도움이 되었다. 또 우리 시대 주택에 보편적으로 필요한 공간은 무엇인지, 프라이버시를 지키기 위해 동원되는 수단이 무엇인지, 구조와 설비는 어떻게 운용해야 하는지 등의 이슈에 대해 답할 수 있었다.

타운하우스도 단독주택인 만큼 마당과의 관계가 중요했다. 거실과 식당, 부엌과 다용도실은 저마다의 마당을 원했다. 또한 효율적인 아파트 평면에 익숙한 이들이 많아서 같은 면적이라도 넓어 보이는 평면을 선호했다. '다양한 마당'과 '효율적 평면'이라는 두 가지 상반된 관점은 '채와 간' 유형과 '단일매스' 유형의 결합을 요구했다. 그 요구는 단일매스 윤곽의 도시형 타운하우스에 내부 마당을 갖게 했고, 채의 집합으로 계획된 전원형 타운하우스에 '단일매스-박스'에서 볼 수 있는 두꺼운 볼륨을 부여했다.

발트하우스 연작 : 전원형 마을-주거

일곱 채의 연작으로 설계된 발트하우스는 가장 큰 규모의 주택이다. 대지의 형상과 조건이 상이한 주택에 일관된 디자인 모티브를 부여하면서도 각 세대가 고유한 특징을 갖도록 계획했다. 단지 전체에 적용된 벽돌을 외장재로 사용하고 공통된 디자인 어휘를 사용했다. 외부 공간의 배치, 처마와 담의 구성, 3층에 마련된

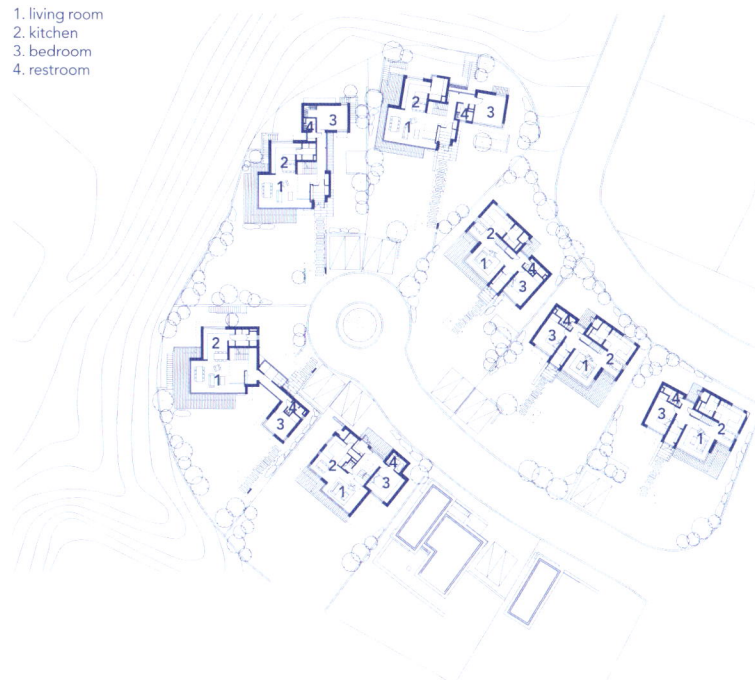

1. living room
2. kitchen
3. bedroom
4. restroom

Fig. 48 발트하우스 평면도 / 효율적인 평면을 위해 하나의 채는 두꺼운 입방체로 만들어 깊이 있는 공간감과 높은 볼륨감, 기능적인 평면을 만들었다.

Waldhaus plan / To provide a more efficient use of area, one of these Chae is made of a thick cube providing a space with a greater sense and higher sense of volume, and creating a more functional area.

Fig. 49 대지의 형상과 조건이 상이한 주택에 일관된 디자인 모티브를 부여하면서도 각 세대가 고유한 특징을 갖도록 계획했다.

It was planned to give a unifying design motif to a building project where the shape of the earth and conditions were different, and also to provide the unique features of this generation.

정자 등이 한옥이 가진 미덕을 반영한 것이라면, 내부 공간의 볼륨감이나 거실과 식당, 부엌으로 이어지는 연속된 공간은 이 시대 생활의 요구를 반영한 것이다. 다양한 마당을 만들기 위해 집을 두 채로 나누었다. 동시에 효율적인 평면을 위해 하나의 채는 두꺼운 입방체로 만들어 깊이 있는 공간감과 높은 볼륨감, 기능적인 평면을 만들었다.

벽돌로 만들어진 분절된 볼륨들은 편안한 스케일을 만들어주고 처마와 담, 발코니 등을 통해 내부 공간과 외부 공간의 사이는 몇 개의 켜로 중첩되며 풍부해졌다. 담장과 조경, 처마가 만들어내는 공간의 켜는 이웃과 적절한 거리를 만들었다. 기후에 대응하는 방식은 돌출된 처마와 창의 배치를 통해 드러난다. 현관 캐노피, 창에 붙어 있는 처마, 낮게 깔린 담장은 집이 기후와 대응하고 외부 세계와 소통하는 방식을 보여준다. 주변 환경과 적극적으로 관계 맺으려는 집의 포즈는 대지 조건이 상이한 각각의 주택에 저절로 고유한 특징을 부여했다. 전원에 집을 원하는 이들은 외부 공간과 내부 공간의 통합된 관계를 원한다. 그것은 본래 모든 집들이 지녔던 미덕이다.

자곡동 타운하우스 : 도시형 마을-주거

서울의 단독주택지는 크게 두 가지 방향으로 변해왔다. 다가구, 다세대 주택이 골목마다 새로 들어서면서 밀도가 지나치게 높아지거나, 아파트로 재개발되면서 기존의 도시 조직이 아예 소멸되는 경우다. 자곡동 타운하우스는 제3의 길을 찾아가는 과정 속에서 만들어졌다.

2007년 자곡동 타운하우스를 통해 제안한 것은 주거밀도를 높이면서도 좋은 환경의 주거지로 업그레이드 하는 개발 방법이다. 100평 내외의 대지에 있던 단독주택을 두 가구를 위한 주택으로 다시 건축한다. 주택의 원소유자는 그 집에 그대로 거주하고 다른 한 집을 분양하는 방식을 통해 공사비와 이윤을 얻는다. 전체 200여 필지 중 일부를 비워 커뮤니티 시설과 공원을 제안했다. 이러한 개발 방식은 기존의 가로와 필지의 형상이 보존될 뿐 아니라 거주자들이 계속 동네에 정

1. entrance
2. living room
3. kitchen
4. bedroom
5. dressing room
6. restroom
7. terrace
8. courtyard

Fig. 50 자곡동 타운하우스 평면도 / 자곡동 타운하우스에서는 단일매스 입방체의 윤곽 안에 중정을 삽입한 도시주택 유형을 제시했다.
Jagok-dong Townhouse plan / The Jagok-dong Townhouse was created so that an inner courtyard is inserted into the shape of an urban residence within the contours of a Single Mass box.

주할 수 있으므로, 커뮤니티와 도시 조직을 유지하면서도 주거밀도를 높이는 장점이 있다.

자곡동 타운하우스에서는 단일매스 입방체의 윤곽 안에 중정을 삽입한 도시주택 유형을 제시했다. 중정을 중심으로 주택의 각 영역을 배치했다. 내부의 모든 영역들이 중정을 매개로 통합되고 이어져 넓은 공간감을 느낄 수 있었다. 또한 마당과 방, 길과 대문으로 이어지는 연속된 공간의 서사를 구축했다. 경사진 대지를 적극 활용하여 실내 주차장을 만들고, 지형의 조건에 어울리는 주택의 단면을 갖도록 했다. 두 세대를 위한 주택이지만 비대칭적인 입면의 구성을 통해 한 채의 집으로 인식되도록 했다. 통일된 재료를 구사해서 하나의 '동네'로 느끼게 했다. 도시주택의 진화는 동네의 진화를 지향한다. 우리가 원하는 것은 살고 싶은 집, 머물고 싶은 마을이다.

Fig. 51 자곡동 타운하우스를 통해 제안한 것은 주거 밀도를 높이면서도 좋은 환경의 주거지로 업그레이드 하는 개발 방법이다.

The Jagok-dong Townhouse was proposed as a development strategy that would allow an increase in per capita residences while still providing an upgraded, quality residential environment.

단독주택 마을 프로젝트

발트하우스와 자곡동 타운하우스를 진행하던 2007년을 전후하여 또 다른 단독주택 마을 프로젝트를 진행했다. 용인 지산, 용인 마평, 제주 중문에 제안한 프로젝트로 마스터플랜과 개별 주택에 대한 유형별 제안이 담겨 있는 프로젝트이다. 경기도와 제주로 이어지던 프로젝트들은 2008년 세계경제위기로 인해 완성되지 못했다. 비록 중간에 진행을 마쳤지만 그 프로젝트를 통해 주택에 대한 생각을 보다 큰 스케일로 넓힐 수 있었다. 길을 만드는 방법, 개별 필지가 분할되는 방식, 공동시설의 구성과 배치 등, 어떻게 마을이 만들어져야 하는가에 대한 생각을 담을 수 있었다.

하지만 이 제안들은 분명한 한계를 갖고 있었다. 어디까지나 판매를 위한 부동산 상품인데다가, 마을이 동시에 조성된다는 것은 우리가 경험했던 마을들과 다른 성격을 지닐 수밖에 없었다. 그렇지만, 주택 공급의 대부분이 건설회사나, 개발회사에 의해 대규모로 분양되는 방식으로 이루어지는 현실을 생각할 때, 적정 규모의 단독주택 마을 개발은 우리 시대에 필요한 대안적 개발 방식의 하나이다. 무분별한 난개발을 막기 위해서는, 마을에 대한 통찰을 바탕으로 한 주택 단지의 체계적인 개발이 요구된다.

Fig. 52 마평동 타운하우스 계획안
Mapyeong-dong Townhouse proposal

Fig. 53 발트하우스 비노 계획안 / 길을 만드는 방법, 개별 필지가 분할되는 방식, 공동시설의 구성과 배치 등, 어떻게 마을이 만들어져야 하는가에 대한 생각을 담을 수 있었다.
Waldhaus vino proposal / I could develop my thoughts on how to make a village, such as how to create streets, the best way to divide individual lots, the structure and placement of community facilities, and so on.

Various Topics of Exploration

If the types of Chae and Kan, and Single Mass – Box are related to spatial systems, it created a new topic of different dimensions to the variety of conditions – social, economic, urban planning, realty development- that exist for our generation. New topics appeared with a variety of residential requests. Demand for detached housing projects increased sharply after 2000, and not only did the number of requests increase but also the complexity of the conditions. This implied that the form of the detached house was becoming more and more divided into a variety of types. If one were to organise these new topics, one might broadly categorise them as the construction of 'Housing in New Town Area' in new cities like Pangyo, and the construction of 'Country House' in rural area 'Town Houses' in new village.

Housing in New-Town Areas: Communality and Individuals

A Purview of Housing in New Town Areas and the Communality of Detached Houses

Urban planning regulations and guidelines for the second generation of new cities like Pangyo did not differ greatly with the first generation of new cities built in 1990s. Residential housing lots were required to have similar size, roads were built on a grid, and the land was made flat. In short, the plans were for bland villages that could not be found anywhere in the world, and not only in Ilsan. The sterile and uninspiring blocks of detached housing stymied any attempt at personalisation. What we find in the villages of new cities is the ultimate expression of thoughtless and inconsiderate development, a veritable zoo of architecture.

As the houses in Pangyo were also subject to the same conditions, the only option was to try and fill each uniform lot of the residential areas with different types (shapes) of housing. Pangyo has stricter regulations for architectural lines that were more advanced in Ilsan, and environmental protections there have fewer regulations regarding the height of buildings and guidelines regarding architectural materials or landscaping were significantly weaker. Furthermore, architectural features are essential for privacy in urban housing, like fences or screens were prohibited. These regulations could only function as intended if applied to areas that were very spacious, much like American suburbia.

In 2008, the clients of eleven blocks in Pangyo met through an online chatroom and then approached me for consultation about planning for construction. Their hope was to have homes that met with their individual tastes and preferences, but also that these homes that would create a pleasing and aesthetically harmonious neighbourhood. Their feeling was that they did not want to repeat the thoughtless development that is so typical of new city residential areas. As the coordinator and architect I suggested a number of guidelines to protect the privacy of each home and ensure the overall harmonious look of the neighbourhood. (Fig. 26) These guidelines had no legal power but they dealt with issues that were not covered by the existing city regulations. A summary of these include height limitations, the inclusion of architectural features that would ensure privacy, ensuring some uniformity in the landscaping of the neighborhood, limiting the types of materials for the exterior walls of the homes. The economic crises meant that the majority of clients were forced to defer construction and landowners who were not a part of the client group that did not follow the guidelines. Consequently, many of our efforts met with failure.

However, through that process I gained an understanding of the client's desire to create a harmonious landscape in their new city neighbourhood, and I tried to remember and apply that understanding in the twenty residential projects I was involved in over the following ten years. That understanding is the public nature of residential homes.

As one element of one part of a city, residential buildings should also be subject to appropriate height and size guidelines, with requirements for appropriate spacing between each house as well as from the road. When height regulations of 6 – 7m for two stories are kept, permission may be given for building an attic or a rooftop room when the occasion requires. To protect the privacy of each home, it is important to be considerate. Windows facing the neighbour's yard should be kept to a minimum and the windows should be covered with a security grille to block views. While the communal personality of the neighbourhood should be established through a consistent range of plant life and trees, individuals are also permitted to add plants according to their personal preference. For exterior materials I consistently chose those that looked as close to the original material as possible: exposed concrete, wood, and natural stones.

When it comes to creating a harmonious village landscape, the goal of defining the communality of residential housing cannot be achieved through eloquent words or passionate speeches. It requires much reflection and, with a vision of the village as context, the construction of a master plan, alongside guidelines that have a deep understanding of the landscape of the village and urban housing, and a mutual respect between the architect and the residents.

Requests per Residential Unit

For our generation, what is required of our homes is a means of self-realisation. Houses are no longer simply a combination of a bedroom, living room and kitchen. It must reflect and function as a space for the interests and hobbies of all the members of the family. This means that within the confines of a single home it is necessary to be able to have multiple spaces that are uniquely different. It follows that even as the architect begins to create the outline of the house, he must also remember to include architectural features that can cater to these different desires and interests. Distinctions and differences have become important goals. One must create a house that groups many different things as one thing. There were many attempts at combining these different things together. By limiting construction materials, producing differences through form and space, creating flexible and multiple shapes under the auspices of a single larger shape, depending on the architectural means of adjusting these distinctions, the features of houses are naturally created.

The most notable characteristic of twenty-first century clients is their access to obtain and share information. Through online chatrooms they can share information about building plans and construction. On the one hand, collecting images of buildings from the internet can be fragmented. The information they gather, or the images they find through surfing the internet can be a useful reference to the architect, but it can also make planning more difficult. This is because these images are often of buildings that are built to very different specifications of land, climate, construction budgets, and are difficult to replicate.

Architects creating buildings for new cities must confront two extremes. The architect's role is to mediate the distance between a piece of land that is utterly lacking in any unique characteristics and the client's desire to live in a personalised and unique space.

The House Nail is one in which the contours of the cube are adapted to suit the shape of the site. (Fig. 28) Three yards are inserted into a site plan that is shaped like a fingernail. The entrance front yard connects the house to the city, and so the entrance and space for the garage are formed. The courtyard in front of the living room - dining room is a central courtyard and so also connects with the kitchen, the dining room, and the bedroom. The small yard between the living room and the stairs creates a view for the living room. These yards are the main attraction of the House Nail. The spaces are developed around these yards and while the landscape is created by facing these yards.

The contours of the exterior walls of the House Nail contrast with the walls that surround the inner courtyard. The exterior walls of the building are finished with black brickwork, and the small windows have a closed off and protected privacy In contrast, the walls of the inner courtyard are made of wood, and the many wide windows give a sense of freedom and shared space. (Fig. 29) By creating a shape that is closed off from the outside but is open on the inside remains private from the surrounding high-rise apartments.

The Pangyo ㄷ–shaped Residence is suggestive of an archetype for urban housing. (Fig. 31) While a garden is often a feature of detached housing, a courtyard is often a feature of urban housing. The courtyard in an urban house is the most efficient way of conceiving of a private exterior space within a small plot of land. The space unfolds, centred on the courtyard, keeping an intimate relationship with the garden. The relationship of the courtyard with the city can be mediated by the opening and closing of a perimeter wall. Architectural apparatus, such as eaves and balconies, windows and walls, terraces and roof gardens, are installed to respond to the climate, providing unique features and organising narratives of experience. Although configured along a single contour, each room enjoys its own exterior space.

Unusually for Pangyo, the land on which **the house Ja-O-Dang** is built is on a slope, and so it must meet the requirements of the land both above and below and it. (Fig. 34) The spatial system for the inside of the house Ja-O-Dang is specified according to the yard, and the perimeter of the house must be constructed in the context of its urban connections. The study for the elder in the house is placed independently in a space between the city and the courtyard. Reflecting the role of the *hanok Sarangchae*, the study has views of both inside and outside the house, and forms a meeting place for the indoors and outdoors. The living room is cosily situated in a deep place. The two courtyards, different in size and character, add serenity to the kitchen and the living room. The three rooms on the second floor have different forms. Although they are all under the same roof, each room has its own unique shape and organisation. This is similar to the relationship between the individual and the family. The home is simultaneously a single house and many houses.

The Country House Experiment: Courtyard, Garden and Room, A Microcosmic Landscape

After 2000, increase in income also resulted in an increase in the number of people who wished to own a home with a garden. The most popular conception was to first own it as a country residence, with a view to retiring there in old age. Unlike urban housing, homes with gardens had plenty of land and beautiful surroundings. The requests were for a suitable division of the outdoor space, and for the outdoor spaces to organically integrate with indoor spaces.

When exploring the placement of the house, and developing the relationship between the indoors and the garden, the *hanok* served as a good precedent. By making careful use of the buildings and walls to divide the large area into smaller areas, I focused on a method that clearly characterised the building's inner spaces such as the *Haengrangchae, Saranghcae, Ahnchae*, and *Jeongja* from the outdoor spaces. The Soswaewon Garden provided a particularly useful frame of reference. (Fig. 37) I learned that one could place the walls and terraces, and placement of the walls along a grid, while at the same time creating a structure that drew in the surrounding landscape. The garden of Chinese architecture also provided a good point of reference. The narrative structures created by narrow and low entryways that led to open gardens made a great impression on me. I recalled the spatial organisation of the Classical Gardens of Suzhou.

The way in which the wall leads from the house to the garden and manages the outdoor areas can also be found in Mies Van der Rohe's Brick House plans, or Frank Lloyd Wright's Usonia Houses. Examples from Modern Architecture show a division of the garden and development, but as a whole, the garden is a background for the main protagonist, the house. In contrast to this, these outdoor areas are clearly defined by their function in a *hanok*. Furthermore, rather than viewing the building as simply another object, it is perceived as a collage that combines with the garden in multiple forms. (Fig. 38) This is because the building exists in a different way to the garden.

As residential projects tend to be relatively smaller in scale, and the indoor spaces must connect with one another within a single building, there are limitations in imitating the *hanok* spacial organisation. However, it is possible to utilise the walls and division of the indoor spaces to create a collage of the building and the garden. The first step in construction is to define the perimeter of the house. Walls are an important way of creating this perimeter. Walls can provide multiple functions beyond simply demarcating the property. It can define the garden and by connecting it to the building it can create new scenes. Embankments or masonry intended to organise the sloping landscape can become a demarcating feature, and when the garden meets a forest, the forest itself can become a demarcating feature. While walls, embankments and woods are appropriately used to mark the perimeter of a property they can help harmonise the property with its surroundings. When the construction of the house has been completed and all these efforts have been expended, it can blend into landscape as yet another part of the view.

The larger the piece of land, the more attention can be paid to creating the entrance. From the parking lot to the entryway garden, from the garden to the hallways, many different spaces are passed through and provide a welcoming aspect. More than anything else, the most important thing is the combination of the indoor space with the outdoor space. By dividing the house into different spaces, different gardens are formed. Country houses that sit on large pieces of land aim at a microcosm, and within that microcosm the many layers of architectural apparatus such as the building and the walls, protects the landscape. Through the layers of combined gardens and rooms, a microcosm intended for residential living is completed.

The Yangpyeong Residence is situated on a comparatively long piece of land and so it was possible to create multiple gardens. (Fig. 39) The living room, dining room, kitchen, master bedroom and guest bedrooms were given as much separation as possible and a garden was placed between each of these rooms. In some ways one might say that the gardens were placed first and the indoor spaces added second.

Although it may look as if the garden is organised on a single layer, there are subtle differences in height between the gardens. In addition to the different heights there are also differences in the materials made in the garden and in the sizes of the garden. The difference in the gardens is defined by their combination with the indoor space. The floor and walls of the dining room garden is finished with the same materials used to the finish the dining room; granite and concrete. The garden that connects with the master bedroom follows the same logic and is finished with wood. The matching gardens and rooms each have unique features and functions. Just as the living room is the focal point of the indoor spaces, the garden connected to the living room is also the central garden. Just as the bedroom must be the most secretive place, so the garden connecting to the bedroom is also the most secretive garden.

The garden is both the background for the house and a main feature of the house. This is similar to how the walls of the house are both a construction method and the background of the house, and is also an independent feature. The building and the space, the garden and the wall, the room and floor, these all exist interdependently.

The Bongga-ri Residence is demonstrative of my recent idea of a 'house composed of houses'. (Fig. 42) That is, a house as a gathering of houses, much like a family unit consists of individual members. Each space in the house comprises an independent 'house'. Between the houses are paths and yards. Each houses embodied within in the house possesses its own space and garden. Although these spaces vary in orientation, size and programmes, a uniform implementation of the same materials and details confer likenesses as in a family. The Bongga-ri Residence thus offers a novel perspective of building an archetypal house, conceived in its gathering of separate houses.

Although there were some drawbacks to building **Gapyeong Residence** on an extreme slope, there are also benefits such as being able to create a parking lot at road level, and that panoramic view provided further distance for the eye to travel, when viewed from the garden or from inside the house. (Fig. 43) The space is developed in an East-West direction so that the southern aspect of the house can be viewed. The underground parking lot has a sunken garden so that natural light can come in via the stairs, and creating an interesting way of entering the living room from the underground garage was an important challenge. When leaving the underground garage one passes the light-filled, sunken garden to proceed up a grand staircase. After passing the extensive garden one comes to the house, spread out in a southwestern direction. The living room and master bedroom are placed in an independent space, and each room enjoys a private garden. Under the sloping roof, a space, finished with red pine, from which one can reflect on the wide landscape, provides a place for meditation and rest. From the rooftop deck, furnished with a hinoki tub, only the blue sky is visible. I envisioned the sky and clouds, wind and rain coming like a gift to greet the client of the house after many years of work.

The Yeoju Residence is my own studio. (Fig. 45) It exemplifies my perspective on a house containing all houses. If a house is a device to exemplify one's existence, ultimately the plurality of existence must render a house not a single house but multiple houses. The Yeoju Residence is a single house but within the cozy area of 40 pyeong (132m^2), contains many spaces that have different surfaces and characters. The bathroom and hallway have a ceiling height of 1.9m and the living room is open to second floor, the 2.3m cubed small room, the study under the sloped roof, the small and narrow attic, in total there are 11 different kinds of rooms that were created from different kinds of needs. Small spaces use stairs built to a smaller scale, furniture and details create a greater sense of space.

    This house transforms itself in a variety of ways. Divided spaces are separated by architectural devices such as sliding doors or windows papered with traditional Korean paper, and help to continue to create multiple changes in this space. This house changes according to the season and the time of day. The eleven different spaces within the house are mediated by architectural devices and the outdoor spaces and every hour takes on a new aspect.

Town Houses and Village Residences: Looking for the Grammar of Collectivity

One new topic of exploration was the residential homes to be developed together as a community. After the mid-2000s developers began in earnest to supply the demand for detached houses. They created packages that combined the sale of land, architectural planning and construction. Clients could reduce the amount of work they had to do, while developers were able to add value to cheaply bought land and sell at a profit. Houses sold in this way were described as townhouses, and where townhouses in Britain or America are like the Korean *yeonrib-jutek* (terraced housing with multiple units), in Korea the term is used to refer to a group of houses developed together at the same time. This shared development was very popular for a time as it cut down on construction costs, and often includes public community facilities and a maintenance office, providing security and convenience.

The construction of a townhouse is based on the premise of shared development, and so it inherently requires that details and architectural materials are uniform. What is more important is the development of the plan. This is because the layout of the inner space of the aforementioned detached housing must be persuasively attractive to a wide majority of clients. Conceptualizing the form of urban housing and my experience of practical application in construction was of great help in developing the appearance of the townhouses. It enabled me to articulate how one might go about dealing with issues such as what the universal requirements of space are for this generation, what strategies are necessary to protect privacy, and how load-bearing and installation must be managed.

Gardens are just as important for townhouses as they are for detached housing. The living and dining room, kitchen and utility rooms all required a garden of their own. Furthermore, the many residents used to the more efficient floor-plan of an apartment preferred wider looking spaces. The diversity of garden types and ergonomic plans were two contrasting demands that required the combination of Chae and Kan with the Single Mass Type to combine. That request resulted in the reproduction of the contours of the indoor garden in the urban townhouse Single Mass Type and the thick volume of the Single Mass – Box, which could be found in the Garden type townhouses that were planned based upon the grouping of Chae.

The Waldhaus Series: Garden Type Village–Residences

The Waldhaus, built as a series of six Residence, is the largest house in the whole complex. (Fig. 49) It was planned to give a unifying design motif to a building project where the shape of the earth and conditions were different, and also to provide the unique features of this generation. The brickwork used throughout the complex is used as exterior materials and uses a unified design language. If the placement of outdoor spaces, the structure of the eaves and embankments, the pavilion provided on the third floor are all aesthetic characteristics of the traditional *hanok*, the indoor space is adapted to modern life in terms of its spatial volume, the connecting spaces between the living room and dining room, and the kitchen. In order to create a variety of gardens the houses are split into two Chae. Simultaneously, to provide a more efficient use of area, one of these Chae is made of a thick cube providing a space with a greater sense and higher sense of volume, and creating a more functional area.

The brickwork division of the space creates a comfortable scale between the eaves, walls and balconies, forming multiple layers between inner and outer spaces, creating a greater sense of plenty. The layer created by the walls, the view and the eaves create an appropriate distance between the neighbours. Strategies for dealing with the climate are manifested in the placement of the eaves and the windows. The canopy of the porch, the eaves attached to the window, the low walls are all intended make the house responsive to the climate and show its method of communication with the outside world. The explicit attempt to create a connection between the house and its surroundings is a natural consequence of each of the lots having unique characteristics. Those who are looking for a house with a garden are actively seeking a relationship between the indoors and the outdoors. This is a positive benefit that originally belonged to all houses.

Jagok-dong Townhouse: Urban Type Village Residences

The detached housing district of Seoul has largely followed two trends of transformation. Multi-unit homes and condominiums have increased, springing up on every street leading to a sharp increase in population per capita, and apartment complexes that have been re-developed and have completely obliterated the existing urban structure. The Jagok-dong Townhouse was created while attempting to find a third way. (Fig. 51)

In 2007 the Jagokdong Townhouse was proposed as a development strategy that would allow an increase in per capita residences while still providing an upgraded, quality residential environment. The detached houses currently existing on the 100 pyeong area (330m^2) were rebuilt as a house containing two units. The original owner would live in one house and sell the other to recoup the cost of construction and create a profit. I proposed emptying some of the 200 lots for community facilities and a park. Through development strategies like this the original streets and lots were not only preserved but the residents could remain in their village, ensuring the preservation of both community and existing urban fabric while increasing the residential population.

The Jagok-dong Townhouse was created so that an inner courtyard is inserted into the shape of an urban residence within the contours of a Single Mass box. Areas within the building were placed in relation to the courtyard. The courtyard functioned as a mediator for all the spaces within the house while creating a greater sense of space. Furthermore, the passages between the garden and the room, the street and the entrance were constructed with a narrative of connecting spaces. By actively using the sloping land, an indoor garage was made, and the house was adapted to suit the plot it was built on. Although it is a house intended for two families, the asymmetric facade of the entrance way suggests a house with a single Chae. The use of unified materials means it feels like a single village. The evolution of the urban house looks to the evolution of the village. What we desire is a house in which we want to live, a village in which we would like to linger.

Detached House Village Project

I worked on other detached house village projects before and after the construction of the Waldhaus and the Jagok-dong Townhouse. As projects proposed for the Yongin Jisan, Yongin Mapyeong, Jeju-do Joongmun, the master plan and the proposed shape of the respective residences are contained in this project. (Fig. 52, 53) In 2008, the international economic crises prevented the completion of the Gyeonggi and Jeju projects, both of which in that year. Although the project ended before completion, it enabled me to expand my ideas about houses on a larger scale. I could develop my thoughts on how to make a village, such as how to create streets, the best way to divide individual lots, the structure and placement of community facilities, and so on.

However, this proposal also clearly has its own limitations. It is, ultimately, a product intended for sale, and the simultaneous creation of the village can only result in a character that is not the village of our experience. However, when considering the realities of the way that the majority of construction companies that produce these houses put them on the market, this development strategy of creating smaller groups of detached housing seems a crucial one. To prevent the unplanned and unregulated urban development that occurs without differentiation, a system of developing housing complexes based on insights into village life is vital.

주택의 변화를
부르는 것들

That which
which
Transforms
Houses

새로운 실험과 진화하는 주택

건축가로 살면서 가장 즐거울 때는 새로운 도전을 할 때이다. 그동안 다양한 종류의 프로젝트를 수행했지만, 특별히 주택 프로젝트에서 새로운 실험을 할 기회가 더 많았다. 아무래도 개인 주택은 자본이나 정치의 논리에서 벗어나, 삶과 공간의 문제에 집중할 수 있기 때문일 것이다. 지금까지 주택설계를 이어오며 실감한 것은, 주택은 오래된 형식이지만 끊임없이 새로워지고 있다는 것이다. 새로운 도전의 계기를 마련해주는 여러 요인들 가운데, '가족-관계', '프로그램-밀도', '테크놀로지', 그리고 '기억'에 주목한다.

Fig. 54 방배동주택은 3대가 함께 사는 대가족의 주택이다.
The Bangbae-dong Residence is a multi-generational home where three generations can live together.

Fig. 55 커뮤니티를 위한 공간을 사이에 두고 부모 세대와 자녀 부부 세대는 최대한 거리를 두게 된다.
Placing a community space in between the parents and the children allows the maximum physical distance between them.

가족-관계

집은 다양한 관계가 맺어지는 사회이다. 가족의 구성은 집집마다 다르고 그 관계 역시 다양하다. 주택은 가족-관계의 인프라이다. 과거에 비해 가족-관계의 스펙트럼이 크게 넓어졌고, 그 결과 주택 공간의 스펙트럼도 넓어진다.

　판교 K 주택의 경우는 은퇴한 부부의 집으로 주말마다 자녀부부와 손자들이 방문한다. 노부부를 위한 집이기도 하지만 자녀와 손자들에게는 주말주택이다. 손자들의 놀이와 안전이 중요한 설계의 관점이 되었다. 방배동주택은 3대가 함께 사는 대가족의 주택이다. 이런 경우 각 세대 간의 독립성은 매우 중요한 요소가 된다. 커뮤니티를 위한 공간을 사이에 두고 부모 세대와 자녀 부부 세대는 최대한 거리를 두게 된다. 손자 세대 영역은 부모 세대와 자녀 부부 세대 사이에 두어 양쪽의 보살핌을 받을 수 있다.

　가족들마다 프라이버시 정도가 다르다는 것도 흥미로웠다. 과천주택의 경우는 얇은 벽과 반투명 유리 칸막이가 말해주듯, 어린아이를 둔 까닭에 가족 간에 프라이버시의 경계가 느슨한 편이였다. 한편 O 주택의 경우는 그 반대로, 장성한 자

1. entrance　3. bedroom　5. restroom　7. parking lot　9. living room
2. kitchen　4. dressing room　6. backyard　8. family room　10. deck

Fig. 56 방배동주택 평면도 / 도로 레벨보다 반 층 정도 높게 조성된 대지 레벨 덕에 스킵플로어 형식의 구성을 취할 수 있었다.
Bangbae-dong Residence plan / Thanks to the level of the site, formed a half-floor higher than the road level, the house came to have a skip-floor form.

That which Transforms Houses

1. entrance
2. living room
3. kitchen
4. bedroom
5. restroom
6. dressing room
7. study room
8. family room

Fig. 57 (위) O 주택 평면도 / O 주택의 경우는 장성한 자녀들의 공간을 프라이버시가 확실히 보장된 영역으로 만들어야 했다.

(top) O Residence / In the instance of the O Residence, privacy was as important for their grown-up children.

녀들의 공간을 프라이버시가 확실히 보장된 영역으로 만들어야 했다. 공간의 관계는 외부 형태에 그대로 투영이 되어 가족의 숫자만큼의 경사지붕이 생겨났다. 이러한 경우는 '집들'로 집이 구성된다는 표현이 어울린다. 봉가리주택이 가장 극단적인 경우로 각각의 방을 쓰는 구성원들의 완전한 독립이 요구되어 방과 화장실은 하나의 세트가 되었다. 결국 집은 '집들'의 집합이 되었다.

부부의 공간적 관계도 생각보다 스펙트럼이 넓었다. 침실을 공유하는 경우부터 층을 각각 다르게 쓰는 경우까지, 요구하는 공간의 조건이 실로 다양했다. 한 사람만을 위한 집도 있었다. 한 사람의 취향을 위해 모든 공간이 만들어졌지만, 그 경우에도 친구와 친지의 방문을 예상해 함께 있는 공간을 꼭 두었다. 인간은 사회적 동물이라는 명제를 주택을 설계하면서 되새기게 되었다.

이렇듯 주택은 다양한 가족-관계의 스펙트럼이 공간에 담긴다. 전통적인 가족의 개념이 흐려지면서 주택은 커다란 변화를 맞고 있다. 가족이 아닌 사람들로 구성된 가족, 1인 가족, 협동조합 가족 등, 새로운 가족의 등장은 새로운 주택을 요구한다. 건축가는 그 최전선에서 변화하는 가족-사회를 공간 안에 구축한다.

프로그램-밀도

도시에 인구와 산업이 집중되면서 개별 필지의 밀도를 올리려는 압박이 가중된다. 경제 논리에 의해 주택이 있던 자리는 높은 빌딩으로 바뀌었다. 근래에 상업시설이나 업무시설 등, 복합용도(mixed-use)를 지닌 건물에 주택이 들어가는 '복합건물-주택'으로 개발되는 경우가 많아졌다. 주택과 다른 기능이 함께 있는 복합기능건물-주택은 오래된 건축의 형식이지만, 근래 프로젝트를 통해 만난 복합건물-주택의 양상은 다양하다.

아카넷, 혜지원 프로젝트의 경우는 출판사의 상층부를 건축주의 주택과 임대를 위한 다가구주택으로 개발했다. 복합건물-다가구주택 프로젝트는 신축에 필요한 재원 마련과 향후의 안정된 임대 수입을 위해 임대용 주거 공간을 자신의 주택과 함께 계획한 예이다. 공동주택의 성격을 갖기 때문에 프라이버시를 지키는

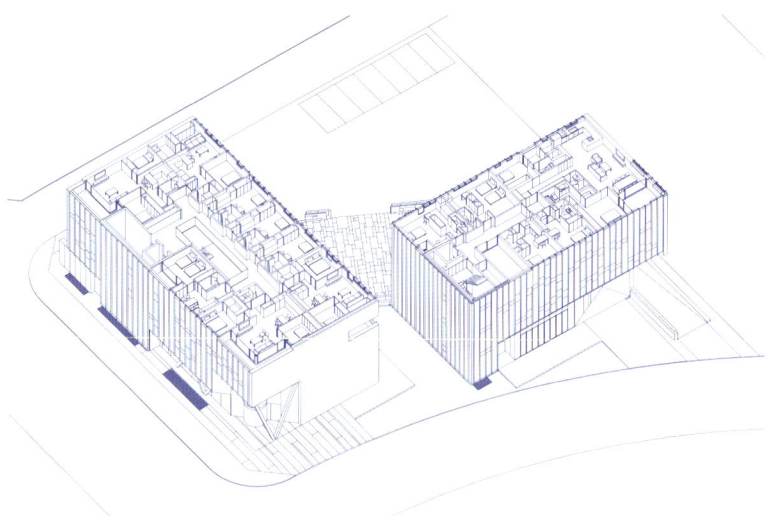

Fig. 58 아카넷, 혜지원 다이어그램 / 복합건물-다가구주택은 신축에 필요한 재원 마련과 향후의 안정된 임대 수입을 위해 임대용 주거 공간을 자신의 주택과 함께 계획한 예이다.
Acanet, Hyejiwon diagram / It is an example of mixed facilities and residential spaces of the building that were developed so that the incomes from the rentals could cover the costs of construction as well as to ensure a future income.

방식, 마당을 공유하는 방법 등이 중요한 관점이 되었다.

그런가 하면, 복합용도 건물에 단독주택을 마련하는 경우도 있다. 청담동 빌딩과 정클리닉, 나루 프로젝트에서는 단독주택이 최상층에 배치된다. 복합건물이 주택을 위한 인공대지가 되는 셈이다. 적층된 대지 위에 만들어지는 복합건물-단독주택 프로젝트는 최상층이 제공하는 전망과 마당을 어떻게 향유할 것인가에 초점이 맞추어진다.

복합건물에 소규모 주택이 삽입되는 경우도 있다. 플레이스 J에는 호텔방 같은 작은 주택이 들어가는데, 손님이나 건축주 자신의 공간으로 활용된다. 소율의 3층에 있는 거주 공간은 같은 층의 작업 공간에 딸려 있다. 건물에 기생하고 있는 듯한 복합건물-소규모 주택은 작지만 특별하게 취급된다. 플레이스 J의 거주 공간에 전망 좋은 발코니가 제공된다면, 소율의 거주 공간은 목재 박스로 제작되어

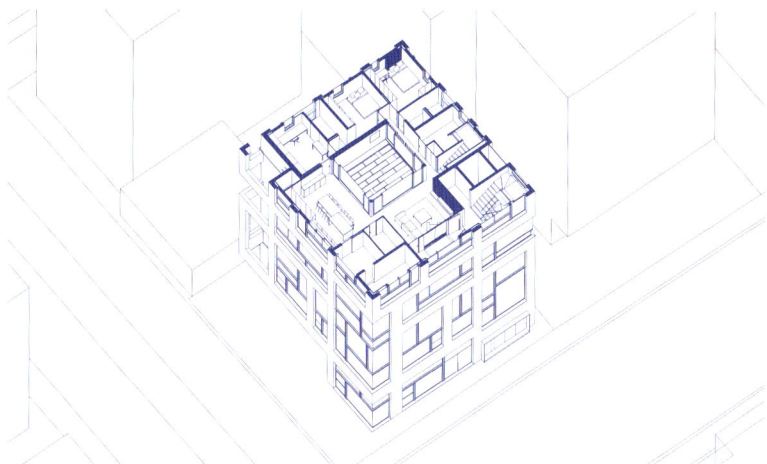

Fig. 59 나루 다이어그램 / 나루는 단독주택이 최상층에 배치된다.
Naru diagram / The Naru project have detached houses on the top storey.

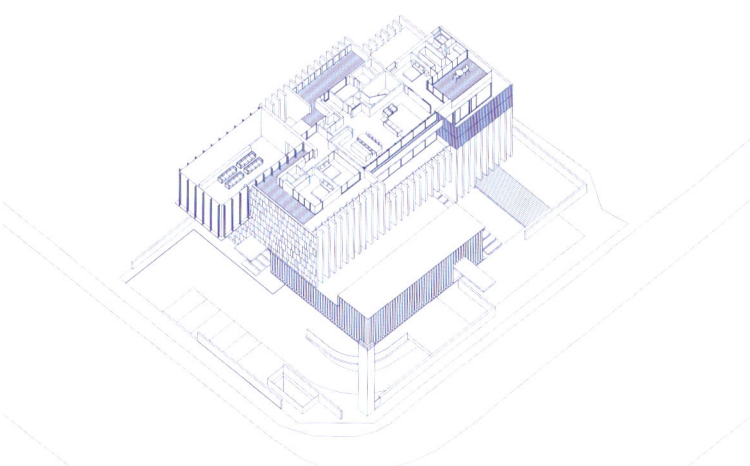

Fig. 60 정클리닉 다이어그램 / 계획설계의 주안점은 클리닉 전문빌딩의 기능을 담아내면서도, 건축주의 거주공간을 최상층에 마련하는 것이었다.
Jung Clinic diagram / The design was focused on how to provide the residential space for client, while performing its function as a building specialized as a clinic.

Fig. 61 (오른쪽) 플레이스 J 5층 평면도, (아래) 플레이스 J 단면도 / 플레이스 J의 거주공간에는 전망 좋은 발코니가 제공된다.
(right) Place J 5F plan, (bottom) Place J section / The residential space in Place J has a terrace with which to take advantage of the wonderful view.

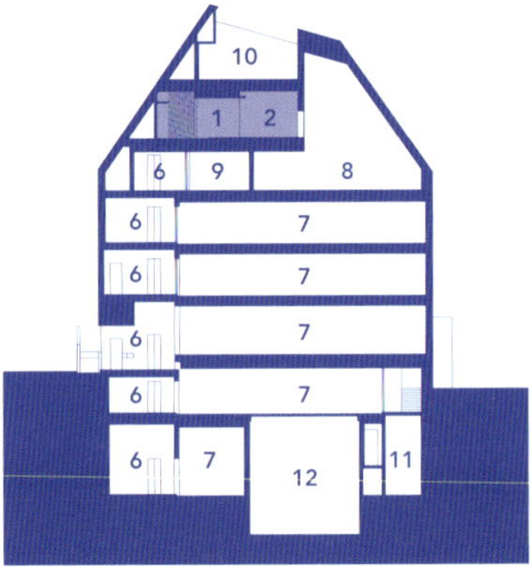

1. entrance
2. living room
3. bedroom
4. restroom
5. terrace
6. hall
7. shop
8. party room
9. kitchen
10. roof garden
11. machine room
12. mechanical parking room

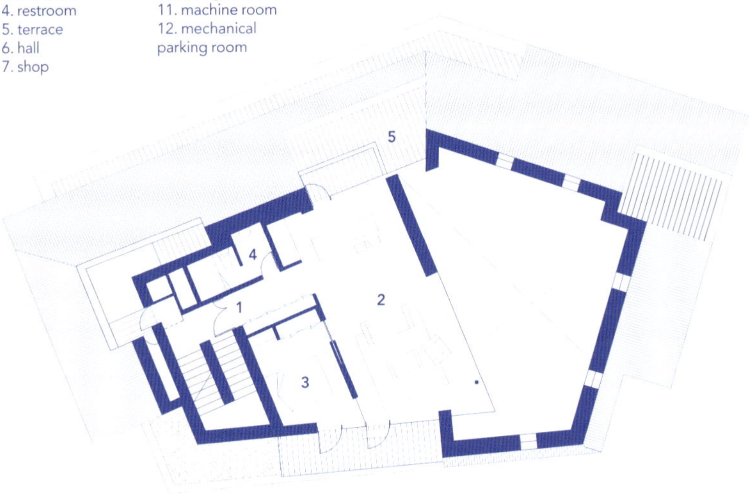

공간 속에 삽입된다.

    복합 기능의 건물과 주택이 함께 지어지는 것은 반가운 일이다. 거주 공간과 일하는 공간이 겹쳐지는 경우가 많기 때문이다. 거주의 영역과 노동의 영역이 가까워질 때, 이동에 필요한 에너지가 절약되어 결과적으로 친환경 도시가 된다. 무엇보다 거주하는 지역이 진정 자신의 마을이 된다. 도시, 지역, 마을에 대한 사랑은 그 장소에서 일하고 거주하는 것에서 출발한다.

Fig. 62 플레이스 J는 개인이 소유한 건물이지만, 동시에 누구에게나 개방된 프로그램을 갖고 있다.
Though Place J is a privately owned building, it is open to the public.

Fig. 63 무엇보다 일과 만남, 놀이가 행복하게 어우러지는 삶을 이곳에 담고 싶었다.
Most of all, I wanted to put the happy life mixed with work, meeting, and play into this space.

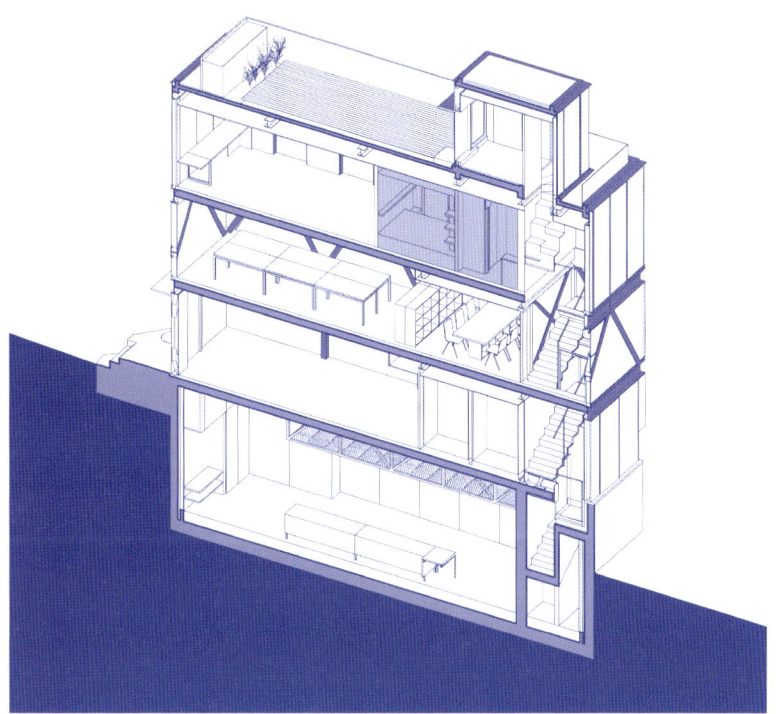

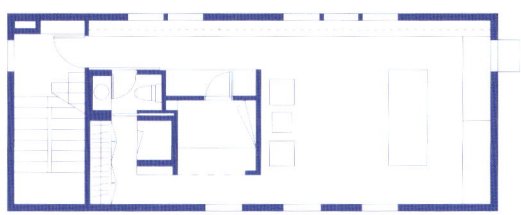

Fig. 64 (위) 소율 단면도, (아래) 소율 평면도 / 단순한 윤곽을 지닌 집이지만, 프로그램과 구조의 질서가 레벨에 따라 다른 만큼 각층의 입면과 내부 공간도 이에 따라 다채롭게 변화한다.

(top) Soyul section, (bottom) Soyul plan / With a simple outline, its elevation and the indoor space of each floor changes in a various ways as each level has a different programme and structural order.

기술

새로운 기술이 개발되고 새로운 제품이 소비되는 가장 중요한 공간이 바로 주택이다. 대개 다른 분야는 개별 제품 단위에서 개발되고 소비되지만, 주택에서는 제품 간 서로 긴밀히 연결되어 있다. 부엌가구와 가전제품의 관계, TV의 크기와 거실은 깊은 관련성이 있다. 특별히 지속가능한 환경이라는 주제는 도덕적 이념인 동시에 새로운 기술의 총아로 부각되고 있다. 지속가능한 환경에 대한 제안은 태

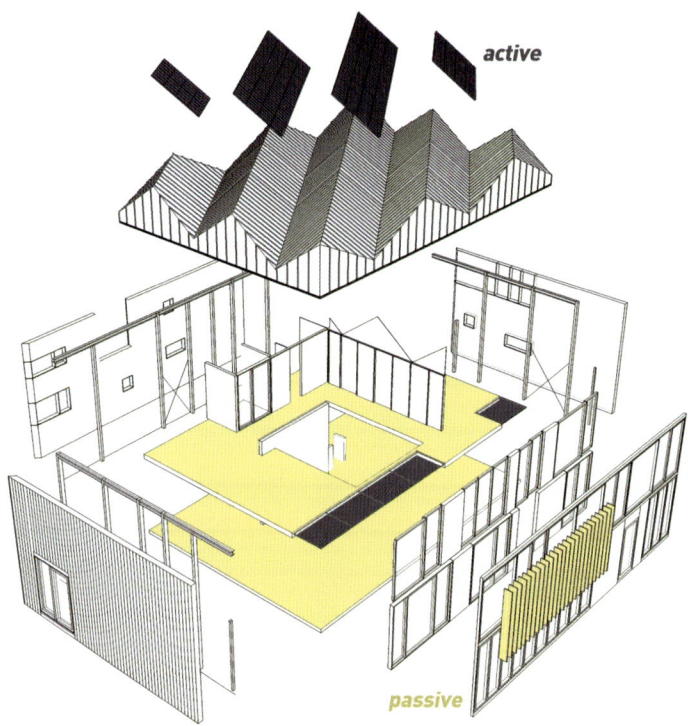

Fig. 65 과천주택을 설계하면서는 건물을 가볍게 만들고자 했다. 최소한의 철골부재를 사용하고, 벽 두께도 최소로 줄였다.
During the construction of the Gwacheon Residence I wanted to make a light house. Using the very minumum of steel framework and subsidiary materials, I also reduced the thickness of the walls.

양광을 도입하고 단열재를 두껍게 쓰는 요소 기술이 아니라, 필지의 개발과 구획, 사회의 시스템, 건축의 규모, 생산과 폐기의 사이클 등 보다 총체적인 관점에서 전개되어야 한다. 개별 프로젝트의 틀 속에서 친환경을 위한 건축의 체계를 제안하는 일은 건축가에게 언제나 도전적인 주제이다.

과천주택을 설계하면서는 건물을 가볍게 만들고자 했다. 최소한의 철골부재를 사용하고, 벽 두께도 최소로 줄였다. 현장 콘크리트 공사를 줄이고, 공장제작 조립부재를 늘렸다. 집을 철거하더라도 폐기물이 적게 나올 뿐 아니라 재활용할 수 있는 재료를 사용했다. 집의 면적을 되도록 작게 하고, 외벽 면적도 줄였다. 태양광 패널을 지붕에 설치하되, 톱니 모양의 형태로 구사해 패널이 외부에서 잘 보이지 않도록 했다.

친환경이라는 목표를 기술적으로 성취하는 데서 더 나아가 미학적인 생성원리로 끌어올리고자 했다. 이는 얇은 두께의 벽체, 프레임을 최소화한 창의 디테일, 지붕의 경사면이 만든 공간감, 경제적이고 소박한 디테일 등으로 구현했다.

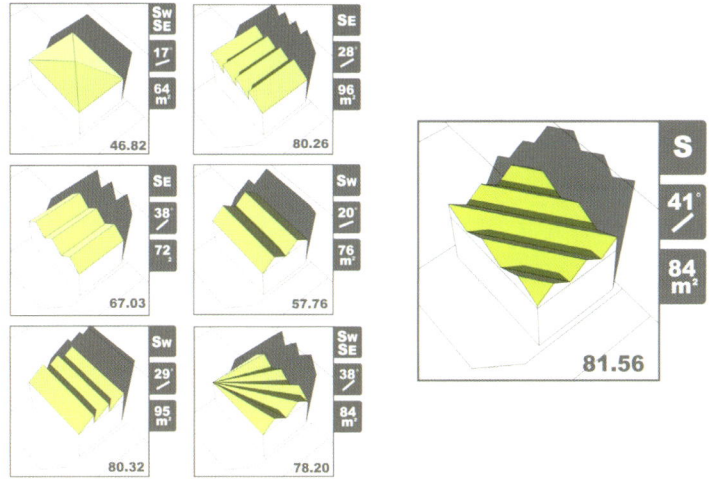

Fig. 66 여러 개로 분절된 경사지붕은 '미학적인 의도'와 '공학적인 필요'의 결과다.
The sloped roof are segmented into several sections and is the result of 'aesthetic intentions' and 'engineering needs.'

각도가 다른 두 개의 질서, 지붕과 벽이 엇갈려 교차하면서 즐겁고 풍요로운 내부 공간이 형성됐다. 건물의 크기에서부터 구조와 설비, 디테일까지 새로운 방식에 도전한 과천주택은 유머러스한 모습이 되었다. 새로운 시스템과 대면하면서도 건축은 얼마든지 유쾌해질 수 있다.

건축에 쓰이는 기술은 현대의 첨단기술보다는 오래된 기술에서 배울 때가 많다. 전통적인 기술 안에는 지역과 기후의 특성이 담겨 있기 때문이다. 습한 여름 때문에 환기를 위한 창호가 많고, 여름의 뜨거운 햇빛을 막고 겨울의 따스한 햇빛을 들이기 위해 처마를 두는 한옥의 건축술은 현대의 주택에 좋은 모범이 된다. 주택은 언제나 오래된 기술과 현재의 기술을 함께 요구한다.

Fig. 67 지붕은 정남 방향으로 경사지붕을 두어 태양전지(solar-cell)를 배치하고 전력을 생산할 수 있도록 계획했다. 지붕 크기와 경사도 등은 태양의 고도와 움직임에 따라 조정되었다.
As the roof looks towards the south, solar-cells are installed on the sloped roof to generate the energy. The size and inclination of the roof were adjusted to the altitude and gradient of the sun.

기억

80년이 된 일본식 가옥 후암동 눅을 설계하게 된 계기는 그 집 아이들이 시멘트로 만들었다는 꽃 벽화였다. 오래된 시간의 기억이 일상의 골목에서 숨 쉬고 있는 듯했다. 눅의 내부 공간에 담겨진 기억을 새롭게 구성하는 일이 매혹적이었다면, 주택과 도시가 만나는 접점을 새롭게 구성하는 일은 도전적이었다. 그것은 서로 다른 시공간을 조율하는 일이기도 했다. 과거의 재료와 우리 시대의 재료를 만나게 하고, 새롭게 만든 처마 아래로 작은 마당을 품었다. 하나의 작은 주택을 통해 80년 동안 쌓인 서로 다른 시간이 새롭게 배열되었다. 벽돌과 시멘트가, 샌드위치 패널과 목재가 처마 아래서 만나 지난 기억을 다시 구성한다. 새롭게 구성된 기억의 공간은 집을 연속된 역사 속에 안착시키고, 그 집에 정체성을 부여하고, 마침내 과거의 기억으로부터 미래의 시간을 향해 나아가게 했다.

Fig. 68 80년이 된 일본식 가옥 눅은 내부 공간에 담겨진 기억을 새롭게 구성하는 일이 매혹적이었다.
The eighty-year old Japanese style house Nook is charming that the idea of creating anew the memory contained in the interior.

Fig. 69 벽돌과 시멘트가, 샌드위치 패널과 목재가 처마 아래서 만나 지난 기억을 다시 구성한다.
Brick and cement, sandwich panels and wooden eaves meet to re-create old memories.

우리 사회의 욕망 저편에, 정체성에 대한 소망이 있다. 잃어버린 정체성을 다시 구축하기 위해 과거를 바라본다. 조선시대의 한양을, 일제강점기의 한성을 오늘의 공간으로 호출한다. 도성을 다시 만들고, 북촌 한옥마을을 보전하는 시도에는 잃어버린 정체성을 과거의 기억으로 대신하려는 열망이 담겨 있다. 그것이 지적 퇴행, 박제된 기억의 포르노가 되지 않으려면, 우리의 도시와 거주 공간을 과거로부터 현재에 이르는 시-공간의 체계로 구축해야 한다. 그 과정을 통해 서로 다른 시간이 서로 이어진 기억의 별자리를 우리에게 선사하고 마침내 우리 도시의 정체성으로 자리 잡게 될 것이다.

Fig. 70 새롭게 구성된 기억의 공간은 집을 연속된 역사 속에 안착시키고, 그 집에 정체성을 부여한다.
This newly created space for memories continues its place in history, and gives the house an identity.

New Experiment and Evolving Housing

An architect is happiest when he is faced with a challenge. Although I participated in a large variety of projects, I was given more opportunities to experiment with residential housing. This is doubtless because a residential home, removed from theoretical issues of resources or fro politics allows one to focus on the business of creating spaces to live in. What I have particularly felt in my experience is that a house might be the oldest form, but it is endlessly being renewed. Of the many opportunities that I was given to challenge myself, I focused on family relationships, programmes and density, technology and memory.

Family – Relationships

A house is a place of many social relationships. The structure of a family is different for each family and their relationships are equally varied. In comparison with the past, the spectrum of family relationships have widened and as a result, spaces created for homes have also widened.

In Pangyo K Residence, the parents and grandchildren visit the home of the retired grandparents every weekend. While it is a home for a retired couple is also a weekend home for their children. Play areas for the grandchildren and associated safety concerns were important elements in the construction of that home. The Bangbae-dong Residence is a multi-generational home where three generations can live together. (Fig. 54) In instances like this, independence for each generation becomes an extremely important element. Placing a community space in between the parents and the children allows the maximum physical distance between them. Space for the grandchildren is also placed between the grandparents' space and the parents' space so that both pairs can help to care for them.

It was also interesting to see the different degree of privacy required by different families. As is apparent in the thin walls and half transparent glass dividers in the Gwacheon Residence, the fact that they have small children makes privacy a more flexible issue for the owners. On the other hand, in the instance of the O Residence, privacy was as important for their grown-up children. (Fig. 57) The spatial relationships within the home imitated that of the outdoor structure and so, there are as many tilted roofs as there are family members. In homes like this, the expression many houses in a single house is especially appropriate. As the most extreme example, the Bongga-ri Residence, the family members required that each person have their own room and en-suite bathroom. Thus, the residence is a combination of multiple houses.

The spatial relationships that exist between couples were also wider in variety. From sharing a bedroom to using entirely different floors, the requests for different spaces were extremely varied. There was even one home intended for a single person. Although it was created with the preferences of a single person in mind, in those cases spaces were created with the expectation of friends or family coming to visit. The aphorism that man is a social animal was brought home to me.

In this way, residences contain a vast spectrum of family relationships. As the traditional idea of a family changes, the house undergoes great changes. Families that include friends as well as family, one person families, co-operative families, the introduction of new families require new houses. The architect is at the frontier of creating spaces for these transforming family – societies.

Density of the Programme

As populations and industry become concentrated in urban centres, the pressure to increase the number of individual lots becomes greater. It follows economic logic that where one had detached houses one would have high-rise buildings. Buildings that combine commercial facilities or office facilities and ergo development of buildings with mixed-use facilities have increased. Although the idea of combining a residential unit with facilities for a different purpose is not a new one, the diversity present in modern mixed-facilitiy projects is great.

In the example of Acanet and Hyejiwon, the upper stories of the publishing office were built as the home of the owner and rental homes. (Fig. 58) It is an example of mixed facilities and residential spaces of the building that were developed so that the incomes from the rentals could cover the costs of construction as well as to ensure a future income. As a shared building, strategies for protecting privacy and sharing gardens were important considerations.

On the other hand, there are also buildings where mixed-facility buildings provide the features of detached housing. The Cheongdam Building, Jung Clinic and the Naru have detached houses on the top storey. (Fig. 60) The mixed-facility building becomes a man-made plot of land on which to construct the detached house. The mixed facility-detached housing projects, built on top stories are focused on maximizing the view and the garden that is available in this particular situation.

There are also instances where small homes are inserted into mixed-facility buildings. In Place J, small hotel rooms like homes are inserted for the use of guests or the client. (Fig. 62) On the third-floor in Soyul is a residential space attached to a studio on the same floor. (Fig. 63) Residences that exist like parasites in mixed-facility buildings are small but are treated in unique ways. The residential space in Place J has a terrace with which to take advantage of the wonderful view, while the residential space in Soyul is constructed in a wooden box which is then inserted into the building.

The creation of residential spaces in mixed-facility is a welcome development. This is because, increasingly, one's residential space is no longer mutually exclusive with one's working space. When the domestic sphere and one's sphere of labour are brought into closer proximity, the less energy is required to move from one place to another and this is also results in a more sustainable city. More than anything else, one's place of residence truly becomes like a village. Devotion to one's city, district, and village start with the places where one lives and works.

Technology

The home is the most important place, where new technology is developed and new products are consumed. In general, these are developed and consumed separately, but there are subtle connections. There are connections between the electronics used in a kitchen and the kitchen furniture, and the size of the TV is deeply connected to the size of the living room. The topic of a sustainable environment is a moral question and also the most important feature of new technology. Proposals for sustainable environment do not promote the idea of using solar energy and thick thermal materials, but must be considered from the wider perspective of the development of building lots and sections, social systems, the size of buildings, production and waste cycles. Proposing new sustainable architectural systems within the framework of individual projects is always a challenging subject for the architect.

During the construction of the Gwacheon Residence I wanted to make a light house. (Fig. 65) Using the very minumum of steel framework and subsidiary materials, I also reduced the thickness of the walls. Not only would there be less waste if the house were to become demolished but the materials were also recyclable. I tried to decrease the surface area of the house and I also tried to minimize the size of the outdoor area. Solar energy panels were installed in the roof and by utilizing a sawtooth shape, the panels were disguised and hidden from sight.

To achieve an environmentally friendly house via the means of technology was one that I tried to solve with aesthetic production principles. This included thinner walls, the window details of a minimized frame, the sense of space created by the angle of a roof's slope. I tried to realise these things in economical and humble details. The rules of two different angles, the exchange that occurs when the roof and the walls meets, these things create a happy and abundant interior space. From the size of the building, to its every detail, the Gwacheon Residence, built with these new methods has a humorous aspect.

The technology used in architecture tends to be older, rather than cutting-edge. Traditional technology tends to cater to the characteristics of the local geography and climate. The many windows provided for the humid summers and the eaves that block the sun of the summer, while in the warmth of the sun during the winter months, are features of the *hanok* that can be good models for modern architecture. Houses always require a combination of old and new technology.

Memory

The catalyst for building the eighty-year old Japanese style house Nook in Huam-dong, is said to be the flower that the children of the house made from cement. (Fig. 68) Memories of older times live and breathe in the streets and alleyways of our daily life. If the idea of creating anew the memory contained in the interior of Nook is charming, creating new points of contact for the city and for the house is a challenge. This is a way of adapting different moments in time and space. I combined the materials of the past with modern and nestled a small garden under the newly made eaves. Through this small house, all the different moments of eighty years could be newly placed. Brick and cement, sandwich panels and wooden eaves meet to re-create old memories. This newly created space for memories continues its place in history, and gives the house an identity, finally allowing it to move from the past towards the future.

The other face of society's ambition is a desire for identity. The spaces of today call out to the *Hanyang* of the *Choseon* era, and the *Hanseong* of the Japanese occupation period. Rebuilding the old palaces, and protecting the old *hanok* villages all contain our desire to replace our lost identity with memories of the past. To avoid this becoming an intellectual retreat or as self-indulgent exercise of regret and victimhood, we must rebuild our cities and residential areas in a system that brings them from the past to the present. Through this process, different moments in time will form the constellations of our memories and finally give our cities an identity with which to place them.

Epilogue

주택은 '삶이 형상화된 공간'이다. 개인의 공간이 지극히 사적인 생활을 담고 있더라도, 그 공간은 언제나 그가 속한 '시대'와 그 시대가 속한 '지역'에 닿아 있다. 역사 속에 등장하는 주택의 형식, 이를테면 한옥, 사합원, 마치야… 이런 이름들은 모두 한 시대, 한 지역의 삶의 형식을 내포하는 표상이다. 삶은 전승된 규범과 현재의 소망이 만나는 지점에서 새롭게 태어난다.

주택이 자아실현의 공간이라 할지라도, 주택은 우리 시대가 부여하는 보편성, 역사와 기억, 현대의 기술과 도시의 조건, 사회적 규범을 담을 수밖에 없다. 주택을 설계하는 건축가는 건축주 그 개인의 소망에 주목하면서 시대가 공유하는 문화적, 지리적 기반 위에서 그의 건축을 관철하게 된다.

보편성에 대한 탐구는 평면의 유형과 공간의 체계에 대한 제안으로 이어졌다. 전형이나 유형이라는 단어는 '새로운 것'을 요구하는 후기산업시대에 유통기간이 지난 관점에 불과할지 모른다. 그럼에도, 학창시절 한옥을 실측하면서 갖게 되었던 소망, 즉 언젠가 건축가로서 우리의 도시한옥과 같이 융통성과 조형성을 가진 우리 시대의 주택을 제안하겠다는 꿈을 이루고 싶었다.

우리 시대의 주택을 탐구하는 과정을 통해 사유의 지평을 넓힐 수 있었다. 삶의 형식, 개인과 집단의 심리, 역사와 환경, 도시계획, 법규, 건축재료, 구조와 시공, 공사비 등 다양한 차원의 질문을 만났다. 주택에 대한 탐구는 건축 전반으로 확장되었다. 일산주택의 유형은 보건소 연작과 학교 연작의 바탕이 되었고, 방배동주택에 사용된 재료의 조합은 교회와 병원에도 적용되었다. 양평주택은 넓은 대지를 다루는 방법론이 되었고, 과천주택은 친환경 건축의 지향점을 만들었다. 이렇듯 주택 프로젝트는 건축 사유의 근간이 되었다.

The house is a place where life materializes. Even if the space for the individual is private in the extreme, that space will always belong to its era, and the region to which that area belongs. The forms of houses that appear in history – the hanok, the Siheyuan, the Machiya - these names all exemplify a single era, a single region's way of life. Life must be re-born with inherited principles and with hope in the present.

Even if the house is a space for self-expression, the house cannot help but be an expression of the universal values of our times, of history and memory, of modern technologies and urban conditions, of social values. The architect explores his architecture by focusing on the desires of his client and by standing on the platform of the culture and geographical values of his time.

An exploration of universality leads to a proposal for the type of plans and spatial systems. archetypes and types may be concepts that are no longer relevant to a contemporary era constantly seeking the new. Nonetheless, I wanted to realise that ambition I nurtured when I surveyed *hanoks* in my student days. In short, one day as an architect I hoped to propose a house for this generation that would have both the flexibility and malleability of form of an urban hanok.

In the process of exploring the houses of our times, we can expand the horizon of architecture. I met a wide range of questions, such as the shape of our lives, the psychology of the group and of the individual, history and the environment, regulations, architectural materials, structure and construction, and construction costs. The exploration of houses expanded to an exploration of architecture. The Ilsan Residence form become the context for the Health Care Center series and School projects, and the combination of materials used in the Bangbae-dong Residence could also be applied to churches and hospitals. The Yangpeong Residence revealed to me a methodology for managing a large piece of land, and the Gwacheon Residence became a model for environmentally friendly sustainable architecture. In this way, the residential projects became a foundation for architectural rationale.

주택을 설계하는 일을 열심히했던 것은 주택이 건축의 기본이라는 믿음 때문만은 아니다. 주택을 설계하는 시간이 언제나 행복했기 때문이다. 주택을 설계하는 목적이 행복한 삶을 영위하는 거주 공간을 만드는 것이기에, 설계하는 시간을 통해 건축주들의 행복을 미리 맛볼 수 있었다. 도면을 통해 거실에서 책을 읽고, 함께 저녁을 먹고, 잠에서 깨어 정원을 바라보았다. 나의 행복한 시간이 고스란히 건축주들의 행복이 되었기를 바라는 마음이다. 한 채의 집이 들어서면서 동네가 더 좋아지고, 도시가 조금 더 아름다워지기를 소망한다. 내 손을 떠난 집들 모두, 함께 하는 가족과 그 이웃들과 더불어 행복하기를 바란다.

더 나은 주택, 더 나은 마을을 이루기 위해서는 우리 도시, 우리 사회에 많은 변화가 필요하다. 건축법규와 도시계획을 수립하는 과정이 변해야 한다. 사회는 다양성을 포용해야 하고, 개인의 삶에 더 많은 여유가 생겨야 한다. 주택은 가장 작은 단위의 건축 형식이지만 그 안에 너무나 큰 과제를 담고 있다. 모두 우리의 행복을 위한 과제이다. 100년 후의 도시와 주택이 궁금하다.

My efforts in the construction of houses have not only been motivated by the belief in the fundamental forms for architecture. This is because I have always felt happy when planning a house. If the goal of designing a house is to create a residential space for a happy life, the architect can experience happiness through the time client spends in planning that space. Through the plan, client reads in the living room, dines on an evening meal together, wakes up from sleep to look out at the garden. It is my belief that moments of happiness transfer intact to the family of the house. It is my hope that through the inclusion of one new house a neighbourhood will improve, and a city become a little more beautiful. In every home I have worked on, the intention is for families and their neighbours to be happy.

To create a better house, a better village, change is needed in our cities and in our society. The process by which building regulations and urban planning are established must change. Society must embrace change, and more leisure time must be made available to each individual. A house is the smallest unit of an architectural form, but it is an extremely large task. It is a task that is necessary to the happiness of us all. I wonder about the cities and houses that will exist in a hundred years from now.

2010
여주주택

2012
판교 ㄷ자집

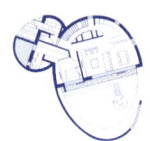
2014
성북동주택

1996
돌체하우스

2010
발트하우스 4

2014
판교주택 B

1997
서초동주택

2017
운중동주택

2009
자곡동 타운하우스

2015
후암동 눅

2016
서래마을주택

2017
구미동주택

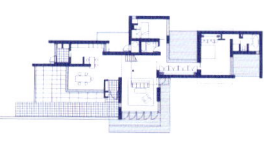
2000
양평주택

2014
O 주택

2014
고양이집

2013
자오당

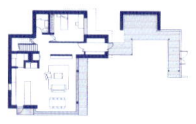
2012
판교주택 F

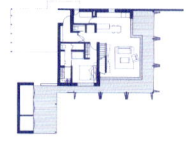
2013
가평주택 아랫집

1995
일산주택

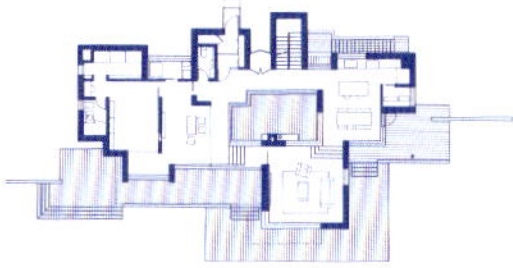

2012 봉가리주택	2010 판교주택 G	2008 S 주택	2011 판교 모서리집
2008 발트하우스1	2008 발트하우스 2	2008 발트하우스 3	2012 손톱집
2006 과천주택	2009 소나무집	2011 이천주택	2010 발트하우스 7
2016 백현동주택	2005 방배동주택	2012 안양주택	2011 판교 ㄱ자집
2009 발트하우스 5	2013 가평주택 윗집	2010 윤교수댁	2010 발트하우스 6

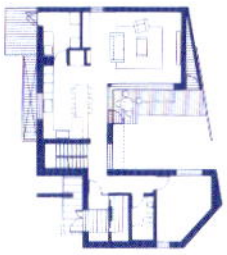

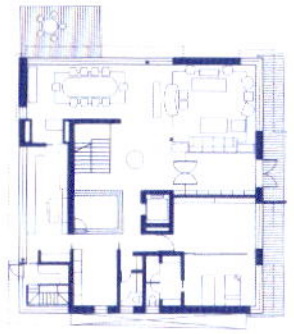

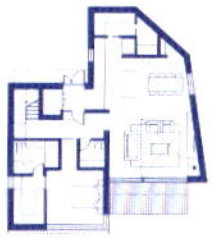

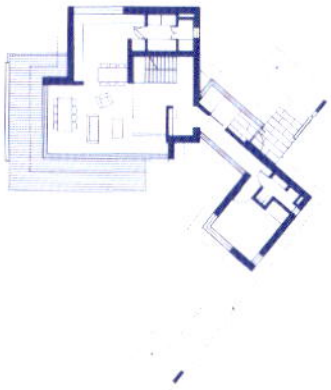

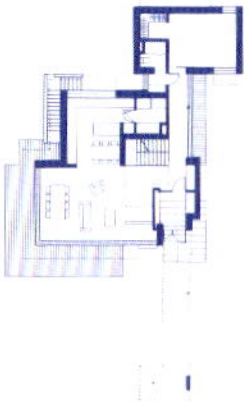

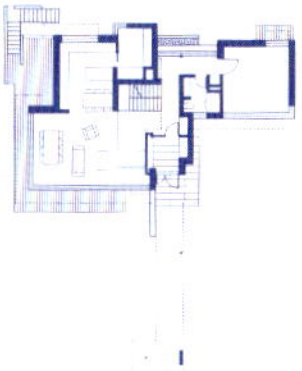

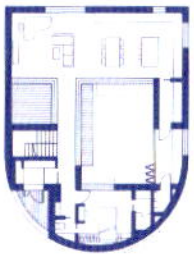

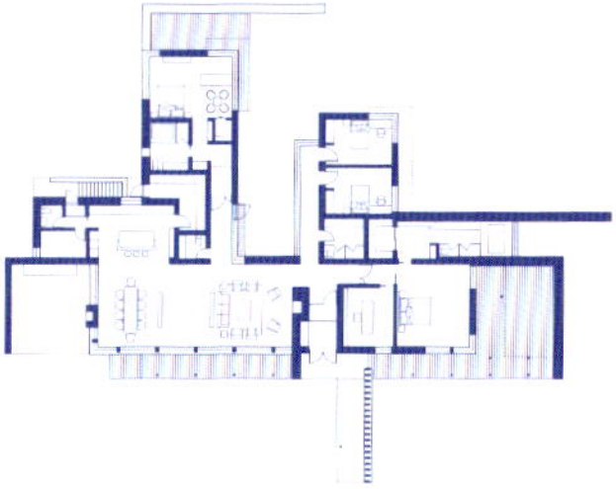

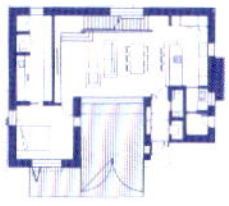

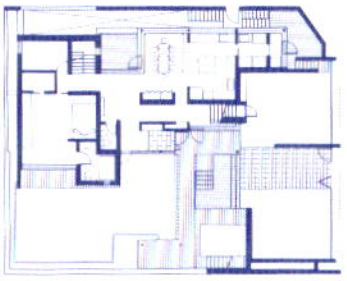

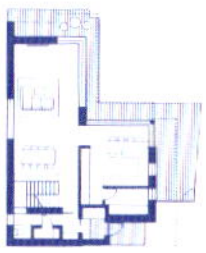

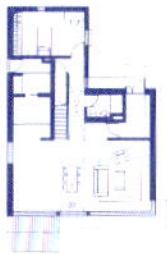

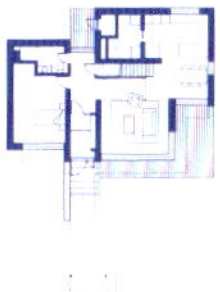

Architect's FRAME 02

주택, 삶의 형식을 찾아서 – 김승회
House, Searching for Life-Forms – Kim Seunghoy

초판 1쇄 인쇄 2017년 8월 29일   초판 1쇄 발행 2017년 9월 1일
FIRST PRINTED 29 August, 2017   FIRST PUBLISHED 1 September, 2017

지은이 김승회 발행인 황용철 편집총괄 박성진 편집 공을채 사진 김재경(별도 표기 외)
디자인 최승태 국문감수 하명란 번역 노성화 영문감수 나탈리 페리스
발행처 (주)CNB미디어 출판등록 1992. 8. 8. (제300-2005-000142호)
주소 03781 서울특별시 서대문구 연희로 52-20 전화 02-396-3359 팩스 02-396-7331
전자우편 editorial@spacem.org 홈페이지 http://www.vmspace.com
ISBN 979-11-87071-15-0 ISBN(세트) 979-11-87071-12-9

AUTHORS Kim Seunghoy PUBLISHER Hwang Yongchul EDITOR-IN-CHIEF Park Sungjin
EDITOR Kong Eulchae PHOTOGRAPHER Kim Jaekyung (unless otherwise indicated)
DESIGN Choi Seungtae KOREA LANGUAGE PROOFREADER Ha Myungran
TRANSLATOR Ro Seonghwa ENGLISH LANGUAGE EDITOR Natalie Ferris
PUBLISHING SPACE BOOKS, an imprint of CNB media
REGISTRATION 1992. 8. 8. (300-2005-000142)
ADDRESS 52-20, Yeonhui-ro, Seodaemun-gu, Seoul, Korea 03781
TEL +82-2-396-3359 FAX +82-2-396-7331 E-MAIL editorial@spacem.org
HOMEPAGE http://www.vmspace.com

©Kim Seunghoy, 2017. Printed in Seoul, Korea

* 파본이나 잘못된 책은 구입처에서 바꾸어 드립니다.
* 이 책은 저작권법에 따라 보호받는 저작물이므로 무단전재와 무단복제를 금지하며, 이 책 내용의
  일부 또는 전부를 이용하려면 반드시 사전에 저작권자와 출판권자의 서면 동의를 받아야 합니다.
* 책값은 뒷표지에 있습니다.
* 이 도서의 국립중앙도서관 출판예정도서목록(CIP)은 서지정보유통지원시스템 홈페이지
  (http://seoji.nl.go.kr)와 국가자료공동목록시스템(http://www.nl.go.kr/kolisnet)에서
  이용하실 수 있습니다.(CIP제어번호: CIP2017021539)

All rights reserved. No part of this publication may be reproduced, stored in a retrieval system, or
transmitted in any form or by any means, electronic, mechanical, photocopying, recording, or
otherwise, without prior consent of the publisher.